Fuzzy Ethics

Fuzzy Ethics

A moral criterion for sustainability

EFRÉN MORENO BENAVIDES

TRANSLATED BY

TANIA MICHELLE TATE

TrueHeart *academic*

– Oxford –

TrueHeart Press (Oxford) Ltd.
Hexagon House
Avenue 4, Station Lane
Witney
Oxford OX28 4BN

www.trueheartpress.com
contact: info@trueheartpress.com

First published in 2013 by TrueHeart Press (Oxford) Ltd.

ISBN: 978-0-9573017-1-9
E-ISBN: 978-0-9573017-6-4

British Library Cataloguing-in-Publication Data: A catalogue record for this book is available from the British Library.

Typeset in the UK by Concerto.

Printed in the UK, USA, and Australia.

Published by TrueHeart Press (Oxford) Ltd., 2013
www.trueheartpress.com

Illustrations by Sonia Lam.

Contents

List of figures

Acknowledgments

I'm grateful to Carmen Segura for her magnificent teachings on the Theory of Truth and her ever-enriching and rigorous reflections; to Sidney Chocron, Marta Jürschik and Silvia Munro for inciting me with their comments to expand upon certain aspects; and to my beloved wife, Carmela, for her countless careful readings and suggestions.

I am especially in debt to my parents, Jesús and María del Carmen, for having instilled in my person the nonconformist restlessness that accompanies those of us who are not perfect. And, of course, to my sisters, Nuria, Xenia and Emma, and family and friends for encouraging my projects and tolerating my absences.

Following the publication of the Spanish book, many friends, colleagues and students enriched me with their comments, reflections and encouragement: I thank you all. I can't list all of you here, but rest assured that you are in my heart and have my deepest gratitude. Special thanks for correcting errors and discussing concepts are extended to Joan Bruno Rodríguez, Luis Lorente, Julián Santiago Prowald, and José Ramón Buenadicha. I would also like to thank Tania Michelle Tate for translating the manuscript at a time that was very difficult for her emotionally, and my editor, George Knott, for accompanying me and helping me to prepare the book with his understanding and good ideas.

Prologue

'Can a group of people isolated on a planet, using only their powers of observation, construct an objective and universal ethic?'

I wrote this essay seeking answers to two questions: 'Is it possible to reach an ethical assessment based solely on observation?' 'If so, is sustainability a foreseeable result of the ethical framework found?' When I began, I did not know where these ideas would take me, but I eventually found the ingredients necessary to outline a fairly satisfactory answer. I hope the effort will be to your liking. Above all, I hope it will be useful to you, the reader, for judging your own position with regard to these questions.

I made four attempts to answer the question 'Can a group of people isolated on a planet, using only their powers of observation, construct an objective and universal ethic?' These can be found under the titles 'Dialogues on life and common sense', 'Reflections of an excursionist', 'Additional content' and 'Sustainability'.[1] The results presented here include definitions of order, happiness, and sustainability that differ from those usually put forward. For example, the definition of sustainability resulting from the ethical framework presented does not explicitly include future generations because their inclusion is a logical necessity. These arguments use a definition of order based on the idea that the greater the number of ways of doing things, the greater the order. Moreover, for the practical application of the ethic, I propose an imperative (a systematic assessment method) that helps to determine which actions, of those actions accessible to our will, are correct from an ethical perspective.

The purpose of 'Dialogues on life and common sense' is to gradually introduce the different questions and answers that frame the search for a universal ethic and the support of an objective moral criterion. In this chapter, I approach the difficulties associated with ethical questions through a debate between the different characters. I try to capture the reader as an active party in the debate, sharing, refuting, and clarifying arguments. The seventeen chapters in the dialogue approach the main ethical questions from the most cosmopolitan position possible. That's why the debate takes place at a campsite, on a single night, between three young people with no special training, who do not know each other, and who discuss examples not far removed from everyday life.

This scenario is intended to provoke fundamental conflicts in the reader. First, ethical questions are just as valid at a campsite as they are at a company or school of philosophy. Second, ethical questions can be addressed within a finite time frame. Third, ethical questions can and should be addressed by normal people. Fourth, the information needed to unravel ethical issues is found in the immediate, everyday world that surrounds us. There is no need to seek obscure definitions, complex arguments, or previously established systems. Observation of the physical world is the fundamental guide that our characters use to construct the ethic proposed in this book. This search leads them to identify a set of ideas that are objective (based on observation), global (applicable to any profession or group of people), and universal (explainable to any person).

The purpose of 'Reflections of an excursionist' is to isolate the main definitions and arguments used in the dialogue. In the twelve reflections comprising this chapter, I isolate the twelve fundamental ideas supporting the ethical framework and moral criterion proposed. These twelve points (which can be viewed as strong or weak, depending on your point of view) were reached using examples and counterexamples in the debate, but I thought it honest not to leave them hidden in the main dialogue. I have uncovered all the pillars supporting my arguments. The idea of this chapter is to provide all the important concepts for anyone seeking to review the proposed system. I have left ideas as bare as possible. This way, anyone who manages to invalidate just one of the twelve points presented can try to destroy the line of argument. For example, someone reading them might discover, as our characters did, that dialogue in the hands of an 'evil' character is a critical element of the system presented.

The chapter 'Additional information' adds nuances to the concepts covered. In particular, it discusses the relationship between the ethical framework presented and change, society, universality, the Universe, and love. Although these dialogues are not necessary for constructing the proposed ethic, they are interesting for exploring its scope.

Finally, the purpose of 'Sustainability' is to relate the proposed ethical framework and moral criterion to sustainability. This chapter uses the terminology of mathematical sets to define concepts such as environment, ways of doing things, and effort. It also considers the relationships between environment, effort, and the moral criterion, and provides an ethical generalisation of the concepts of order and happiness.

Using the proposed moral criterion, I prove, among other things, that allowing the extinction of a species is ethically incorrect. It is a fundamental objective of this chapter to show the conflicts that come about when the criterion is applied. In particular, the chapter focuses on scarcity, interference, environmental, assessment, extinction, finiteness, and sustainability conflicts. I devote special attention to the finiteness conflict because it helps explain why practical application of the moral criterion is necessarily imprecise: practical application of the proposed universal ethic proves it to be a fuzzy ethic. However, the correct application of the criterion and the reduction of conflicts decrease this fuzziness, and make practical application increasingly precise. The chapter also shows how the elimination of conflicts, particularly the finiteness conflict, helps explain why the planet (Earth, in our case) and the beings living on it have intrinsic ethical value.

Finally, we propose a new definition of sustainability. This definition is based on the proposed ethical framework. We conclude that the sustainability conflict is the most complicated to eliminate because it is inherent to the definition of sustainability. We can only aspire to minimise the conflict, as sustainability is an ideal to which the proposed ethic leads.

Note

1 This book is a translation and extended edition of a printed book published in Spanish by Éride Ediciones in 2008, and in digital format by Luarna in 2009. All chapters except 'Sustainability' have been translated from Spanish. (Only Chapter 6 of 'Dialogues on life and common sense' contains any significant variations from the Spanish edition.) The chapter 'Sustainability' replaces a dictionary of terms published in the Spanish book.

Part I

Dialogue on life and common sense

— Dad, what is Paradise?

— I don't know, son. Sometimes I think that, if Paradise exists, it should be a Universe capable of ordering itself on its own...

— This one does.

— Partly, only partly...

1

The defining problem for human beings

The ethic must be objective: based on observation.

The ethic must be global: independent of particular professions.

The ethic must be universal: explainable to all human beings.

It was a special night. The summer heat, not yet suffocating, was extremely pleasant. At last, the atmosphere, utterly clear, made it possible to see the scatterings of stars.

On this night, Hermes was lying face up. He was alone, but unworried by this. He'd had the opportunity to make great friendships in his life, and was fortunate enough to have countless people who supported him. However, on this occasion he was enjoying the contemplation and small sounds of the night without any company.

The moment was particularly awe-inspiring. In that field, solitude shrank his heart to a feeling of near insignificance. It seemed only he could feel the quality of this special night, but if asked why, he would have replied that he did not yet know the reason.

Not long ago, the dreaded 'Year 2000 problem' had finally arrived and passed by, and ever since, humankind had been hurtling forward; rushing into the new millennium with the same indifference it had shown to the conclusion of the previous. A change of millennium seemed a sufficiently exceptional event to strike fear into any intrepid heart, increasing that feeling of infinite smallness and powerlessness that sometimes grips the souls of men. Hermes thought that each and every one of the people who have lived on Earth have probably had the exact same feeling.

About two thousand five hundred years had passed since the glorious age of the Greek philosophers, but if you examined the newspapers and news programmes of today, you would observe no apparent changes in humankind: wars, murders, humiliation, racism, injustice, individualism, separatism, arrogance, terrorism, violence, and other evils were still present on our planet. Humankind's natural surroundings had not changed very much either: diseases, hardships, accidents, earthquakes, floods, volcanic eruptions, droughts, and all manner of catastrophes and unforeseen circumstances continued their plagues.

Hermes wondered just what had changed in those two and a half millennia. Knowledge certainly seemed to have increased: today, every professional does a better job than their counterpart two thousand five hundred years ago. Doctors, engineers, architects, physicists, chemists, biologists, and mechanics had achieved considerable progress in relatively short time. However, psychologists, philosophers, politicians, sociologists, and theologians had made no such spectacular progress. The great questions of humankind remain the same, with no convincing answers for humanity as a whole. Multiple answers to the same question, founded and unfounded, had led to different experiences, religions, and political ideas, prompting humankind to remain divided into groups.

To ensure their continued existence, such differentiated groups must prove all other ideas as erroneous. The powerlessness felt by each member of the group when attempting to successfully prove their ideas, causes despair and loss of patience, so that they also miss the chance to convince or be convinced. This eventually results in isolation or confrontation. A sad fate for humankind, and a situation likely to continue for another thousand years.

Hermes was still lost in thought. He neither knew nor cared how much time might have passed. The distant sound of a mountain stream and the murmur of the wind in the underbrush made him think that the night would never end. In the infinite Universe splayed visibly above him, the moon was notably absent. This was just as well because the sky of stars was revealed in all its splendour, with even the last star visible.

With eyes already accustomed to darkness, he gazed at the soft movements of shadows, and tried to remember their daytime colours. He shone his torch into the darkness, but at that distance the slanting beam was not bright enough. Imagining the possible nuances of vegetation, he found himself gazing again at the horizon, now peppered with endless dots of light. How strange and faraway they seemed when imagined as giant beings of incandescent gas! 'Do I remember the names of any celestial bodies besides the North Star? What happened to the knowledge I acquired in Astronomy?', he wondered, almost out loud, completely absorbed in contemplation of those tiny points, made bright by minute colour changes. How easy it was to forget; how tortuous to remember!

Another thought flashed into his mind. It was late June, and soon it would be his birthday. He felt no emotion towards this anniversary. Supposedly, it would mark the end of a significant portion of his life: with today's life expectancy (one thing which had fortunately changed), a third or quarter, give or take. During this period, he had met many of the objectives imposed by society: he had a good job and the possibility of a good future. His individual position was simple to analyse, but placing himself in his surroundings was more difficult. How could he know whether he was on the right path? Even though he apparently was, this was simply an opinion: his opinion. What might others say?

Around him, the prevailing philosophies of life were 'live to work' or 'work to live', and every degree between. Enjoyment of life. From this train of thought, he could envision some of his friends consuming party nights, daytrips and journeys, without seeming to consider anything more. But life without moderation led to an uncomplicated superficiality. There was a contrast that made him feel uncomfortable. Every day, somewhere nearby, he saw incomprehension, authoritarianism, insensitivity; situations that ended up causing in him a complicated feeling of powerlessness.

It wasn't easy watching the days go by while he, an insignificant part of the world, did not know what to say or do to avoid such behaviour in others and, he might as well admit it, in himself. Obviously, he knew how difficult it was to avoid arguing when your pride has been hurt. He knew how hard it was to keep opinion from easily prevailing over reason. He knew how easy to judge frivolously, even when you knew you lacked the elements necessary to catalogue a particular action as an injustice. It was like trying to fight against some higher force that drags you inexorably toward such actions.

Actions that seem to form the easiest and shortest path because of the natural way in which they germinate inside our heads. However, if we analyse their results with care, these are the most destructive, perhaps as damaging as physical disease the ones that transform themselves, some disappearing while others are engendered as new in their day as those now extinct were in theirs. Those that have been eradicated, few as they are, were eliminated due to the continuous efforts of human beings. But in any case, such ways of behaving have been in the past, and are to this night, travelling companions for many people.

What an ever-changing flow of thoughts! What strong differences between all those ideas he had examined in just one instant, almost without noticing. And, at the same time, what similarity. They all seemed to be joined, connected. Otherwise, why would his intellect draw them to the surface, one after another? It was obvious that they were intermingled: their different nuances were like a palette of colours where a small and continuous change leads almost effortlessly from one to another. What was the imperceptible link remaining completely separate, yet still joining together his ideas, while also forming a part of them? Is there such a fine thread, steeped in the very essence of that which it touches, and impossible to separate from that which it joins?

In this wellspring of understanding which had just now begun to flow up from within, he was interrupted by someone who at first struck him as simply strange. Whoever he was, he did not seem to be a local. The absence of a moon only served to magnify this unique character's original appearance. But Hermes couldn't guess anything more about him, except that he was probably some flamboyant excursionist.

Hermes shone his torch on the newcomer's face…

STRANGER.— Hello. This is undoubtedly a night that invites reflection.

HERMES.— It is. I myself was trapped in my own thoughts when you arrived.

STRANGER.— Sorry if I disturbed you. I couldn't get to sleep, and from my tent I was able to observe the magnificent place you found to abandon yourself into the arms of Morpheus. Since there aren't many of us in the camping area, I was wondering if there would also be room for me. Or perhaps you'd rather be alone.

HERMES—I'm not actually seeking solitude or trying to sleep. The problem I'm pondering is of such magnitude that I'm afraid I'll get lost or despair if I try it on my own. If you like, I would prefer to engage in a dialogue.

STRANGER.— How odd. I thought you were trying to rest under the starry sky. What is this problem you're pondering?

HERMES.— I was just telling myself that humanity hasn't progressed as far as it sometimes seems. Many of the evils engendered by people are still with us today. Humankind is not working towards common objectives, but rather strengthening its differences. And given the complexity of our environment, I find this paradoxical. The trend is obviously to unite, but globally and individually, that connection doesn't quite take off. I think the question could be formulated as follows: Is it possible to create a unifying link between all human beings, regardless of their religion, politics, social status, education, nationality, profession, sex, race, or sexuality?

STRANGER.— My God! Isn't there a simpler problem? One we can solve? I wouldn't like to end up with a headache on such a marvellous night.

HERMES.— Sorry. I know that must have been a bit of a shock. It's not usual to state a problem so openly and directly, especially when we haven't even introduced ourselves. I'm Hermes.

STRANGER.— I'm Xargos. I set up my tent a few days before you, and I noticed that you spend long nights lying out in the open, in this area where the grass is most inviting. I was starting to feel envious! But I really didn't imagine that you were pondering such questions.

HERMES.— (Smiling.) Envy, another evil easy to instil and difficult to eliminate... (Murmuring.) There is a way, a solution!

XARGOS.— What?

HERMES.— Again, I apologise. I like to reflect, and if I have company, that's even better. I'm afraid I've wasted time neglecting these kinds of problems most of my life. Now haste has accelerated my thoughts; I feel I can find a solution.

XARGOS.— Haste is usually a poor counsellor, and even more so if your desire is to find an answer to the dilemma you're considering. If you achieve that, I would consider it a great feat. Wouldn't it be better to start with something simpler?

HERMES.— A problem as important as the one I've proposed has a solution. That's my feeling. And I hope it will eventually become my opinion. If we settled on a simpler approach, we might reach a conclusion more quickly, but the big question we would have left unsolved would accompany me daily, and would make this feeling of time slipping by more acute. So, I've decided that this will be my plan for tonight. If you would join me and discuss these questions, I would enjoy the company.

XARGOS.— I'm not usually a fan of extensive dialogue. And I don't let myself get dragged into discussions where the positions are stated as acts of faith and later disguised with reasoning. So, I should warn you: if that's what you intend to do, I'd rather leave you alone. I don't really feel like wasting time. For me, talking for the sake of talking, when I could just be watching the sky, is the biggest waste of time there is.

HERMES.— I don't like to state positions without first letting them mature either, at least as far as time permits. In that sense, you don't have to worry, as long as you also do as you demand.

XARGOS.— Well, it looks like we both like to reason. If you like, we can try it together.

HERMES.— Of course. My purpose is to start with the question I formulated earlier, and then elaborate on a series of clues that are going through my mind in order to form a convincing answer. I'm not completely sure whether there's a solution to the problem I've formulated. However, I sense that there is.

XARGOS.— If there is, it would be a pleasure to discover. In any case, just finding myself with someone who feels capable of confronting such a question strongly appeals to me. My curiosity has got the better of me. I don't know how you're going to begin such a thread of reasoning, or even how you're going to approach the question. Nevertheless, I do believe that I understand the problem you're addressing. If I'm not mistaken, you intend to seek a link between all human beings that would keep them united with unquestionable solidity, so that no member of the human community would want to go against the community as a whole, or against another member.

HERMES.— That's right. And I intend to do so as simply as possible, and in the most cosmopolitan manner possible. Only then will it be within the reach of everyone. To achieve this, the tools we use must be as universal as possible. What do you think of using only a good dictionary and our enthusiasm?

XARGOS.— That sounds like a good choice: the dictionary is consistent with itself, and we certainly have enough enthusiasm. But the link that we seek is not in the dictionary! If it exists, it's within human beings themselves, who must understand and promote it. The existence of multiple interests is proof that such a link is difficult to find. The possibility of humans behaving in unison as a whole is highly improbable.

HERMES.— Of course, but what they must agree on are the essentials. Even if their particular interests are different, there must be strong cohesion regarding the basic rules of behaviour. How can people, in the absence of Religion and Law, realise that they must not kill? These basic rules are imposed on society, some by law, others by custom, and others by general principles obtained from abstractions made over the generations. If a man or woman is deprived of all this information, can he or she deduce it again quickly and use it throughout life? Can a criterion be found to compare different rules and establish a continuous improvement process?

XARGOS.— Only if he devoted sufficient time would he find the principles that he might be capable of producing by abstraction. And if God sent that man his commandments...

HERMES.— Today, every culture and every society imposes different norms: *carpe diem*; do unto others what you would have them do unto you; engage in dialogue; thou shalt not kill. It would be much better if everyone were to deduce them. If it were possible and they were universally valid.

XARGOS.— Some clearly are, but others are not, and those in between create doubt and anxiety in certain situations.

HERMES.— Aren't you under the impression that human beings are divided into groups because of the differences I noted earlier when posing the question? Given that many of these differences are arbitrary or fortuitous, it seems ridiculous that human beings cling on and try to impose their differences on the rest of humankind. Moreover, I believe people have always tried to achieve cohesion by imposing nationalities, religions, political ideas, and other such conditions. Individually they seek to satisfy their particular interests and concerns. I find it odd that there is nothing higher that can achieve union between everyone. Does it not surprise you that there is nothing external to humankind that binds together our interests universally?

XARGOS.— There is God. God is love, and He is within every one of us. That is the link we are seeking.

HERMES.— At this moment, I do not doubt the existence of God, but neither do I doubt his non-existence. Even so, it is obvious that what you just said is truly beautiful. For two thousand years, God has been consciously present for a part of humankind, and thanks to Him, great progress has been achieved in the mutual understanding of each human being. However, I'm sure you will also agree that atrocious acts have been carried out against humanity in the name of God.

And we must not forget the diverse religious beliefs that exist, and the fact that Christianity alone has been unable to convince the believers of other religions to abandon their ideals.

XARGOS.— That's true. The Church, like most religions, has sought to impose its beliefs on those who did not share them. But the question is not whether the Church has acted rightly or wrongly, but whether God exists and is the link we seek. I believe so because I have experienced the existence of God within myself. I have internal evidence, irrefutable to me, that God exists. This Faith in Him makes me feel good, and I lose my feelings of powerlessness; it enables me to live in peace with myself and with others.

HERMES.— I am truly happy to hear that. But I do not enjoy the good fortune you do, and have not experienced anything within myself that causes me to have as much faith as you have. And until that happens, if you will allow me, I'd like to share a little piece of reasoning I once read. As you know, my name, Hermes, is also the name of a Greek god who was a protector of commerce and all manner of appropriations, legitimate and illegitimate, and a guide to travellers. Son of Zeus and Maia, his astute and clever nature prompted his father to make Hermes his own messenger, and that of the gods of the underworld. Hermes was also in charge of guiding souls to the underworld. I would now like to refer to this last occupation. Of the religions that affirm the existence of hell, it is not difficult to find at least two that regard entry into hell as a punishment for not believing in that religion. If we also consider that such religions regard each other as mutually exclusive, making it impossible to believe in two religions at the same time, we must conclude that all souls must go to hell. That would make Hermes the god with the largest workload, since he would have to guide each and every soul to hell.

XARGOS.— I thought you took this conversation seriously, but I can see that you're treating it as more of a joke. This observation is curious and rather funny, but God understands nothing but love, and right now I don't believe that hell exists. So, this discussion doesn't make much sense.

HERMES.— I believe that it does make sense. At no time have I questioned the existence of God, who exists from the very moment that you believe in him, and ceases to exist from the very moment I deny his existence. I was only trying to express the multitude of ways of thinking and feeling that one can have towards him. You yourself expressed your belief that hell does not exist, even though the Catholic Church claims that it does. I suppose that the existence of hell is a subject that makes you feel uncomfortable; it causes you to lose the peace that God gives you. If I've understood you correctly, God is love and forgiveness for you, and he makes you feel secure.

XARGOS.— That's right. Hell fills me with terror, and I don't believe that God, who is love, could want to terrorise us in that manner.

HERMES.— If I brought up the subject of hell and Hermes, it was to show you that basing our positions on our inner spiritual life leads to a wide variety of intellectual nuances. That kind of subjectivity makes it hard for me to believe that this can be the right path.

XARGOS.— However, I think we should continue on this path. We don't know what itinerary to follow, so ruling out religion as soon as we start seems very risky to me.

HERMES.— To my way of thinking, it's not risky at all. It is essential to leave religion behind. We are seeking a link, but curiously enough, just exchanging a few words has led us to discuss the existence of hell and whether God is love. This situation proves that religion is not the best beacon to follow on our route because you will either get me to believe exactly what you believe, or we will seek another way out by reasoning.

XARGOS.— I repeat that I think it's a bad idea to abandon this path. For me, it is obviously the right one.

HERMES.— On the other hand, I don't think it's prudent to lose sight of our objective by entering into theological discussions that might get out of hand. I think time has proven religion incapable of unifying humankind: new religious movements appear every day, more often than they disappear. At first glance, this doesn't seem to be a unifying path, as the trend is towards a religion customised by and for each person. No matter how religious you are, you can try to seek this bond, link, or connection without religion. If it exists and we find it, we can be sure that it won't be incompatible with any religion. If incompatible, it obviously wouldn't be able to achieve our proposed objective of unifying humankind independently of religion.

XARGOS.— So, you can assure me that we'll manage to unify humankind while maintaining our beliefs?

HERMES.— If you remember, that was the question I formulated earlier.

XARGOS.— Fantastic! This is quite an ambitious problem we'll be tackling. And right now I think we're headed for a most spectacular failure, since there will always be people with political or spiritual ideas one might describe as malignant, who would prevent us from reconciling some people with others.

HERMES.— That problem could certainly become an insurmountable obstacle, but let's set that matter aside and try to find a solution later. If we don't find a unifying link, then why worry about possible threats to it? A better time to reconsider that question might be at the ending of our path, if we do in fact reach a conclusion.

XARGOS.— How should we approach our search for this connection, this link between all human beings? If it can't be a religious link, then it certainly can't be a political, nationalistic, or ethnic bond. And it can't be a professional, educational, or social link. I don't see a way forward.

2

On common sense

Common sense enables us to judge that isolated things tend to become disordered.

This common sense is objective, global and universal.

HERMES.— If we consider a group of people of any ideology, creed, race or profession...what would they be obligated to share?

XARGOS.— I guess it would be the environment surrounding them, if the environment were the same for the whole group. And even if identical, it wouldn't be the same. The host of sensations, emotions, and reflections it would generate in each person would differ. It is well known that each being perceives the sensory world differently, meaning that the same reality can be interpreted differently by each person.

HERMES.— That's true. But is there any characteristic in the entire sensory world identical for all persons, even if their five senses and their psychology cause them to perceive reality in different ways? I believe there is. Isn't there something that we all refer to as *common sense*, that we assume to be the heritage of all reasonable persons? And don't we appeal to this common sense in all life situations where we seek consensus, whether they be professional, personal, or religious?

XARGOS.— Yes, but we both know that what we call common sense is usually the least common of all the senses.

HERMES.— What do you mean?

XARGOS.— It's easy to see that different people faced with the same problem will come up with different solutions while acting in accordance with common sense. It is therefore not a sense common to all, but one different for everyone.

HERMES.— Let's be careful to avoid such distinctions, while adhering only to the essential. We've started a dialogue on common sense, and it's possible that each of us has a different definition in mind. I think a good definition might be the ability that most people have to judge things reasonably.

XARGOS.— Taking the dictionary definition is undoubtedly a good way to get our ideas straight. But if you select another meaning, you'll find that common sense is the inner ability to receive and imprint all types and images of objects sent by the outer senses. In other words, common sense collects the specific sensations provided by the other senses and combines them. The result is sensory knowledge of objects, as opposed to mere sensations.

HERMES.— I can certainly see that you're willing to make this dialogue difficult. Let's choose the definition best suited to our discussion. In order for the judgment issued by common sense to be reasonable, the information received by the senses must also be reasonable. If you'll allow me to change my mind, I would like to choose your definition as the most suitable. Mine, in addition to requiring prior sensory information, assumes that a judgment must be made. From now on, let's define *common sense* as you suggested earlier, if you agree.

XARGOS.— I agree, and for the sake of clarity, I'd like to repeat the definition we've adopted: *Common sense is the sense that collects the specific sensations provided by the other senses and combines them, resulting in sensory knowledge of objects as opposed to mere sensations.* In this way, common sense establishes the sensory data upon which *the ability to judge reasonably* will act on the elements that also constitute common sense, according to your definition. Because one must first know in order to judge, my definition must obviously precede yours. However, at no time does it exclude yours; it simply comes first.

HERMES.— I agree completely. Let's place your definition first, and mine second. As you said, mine needs yours in order to exist.

XARGOS.— How shall we continue? Even though we've agreed on a definition satisfactory to both of us, we still have the added difficulty of determining whether this kind of common sense is different for each person.

HERMES.— Because there would be nothing to discuss if it were the same, we have to assume that it is indeed different, meaning that sensory knowledge of an object, which we've said is produced by common sense, would be different from one person to another. But wouldn't an individual who acquires such sensory knowledge over time reach the same conclusion as another, even if the objects of their knowledge were different? Regardless of who the observers are, wouldn't they end up obtaining some small portion of knowledge identical for all of them?

XARGOS.— I don't quite understand. Please explain.

HERMES.— If we were to ask a large number of people what would happen if we neglect a garden, drop fine china on the floor, or leave food unrefrigerated, wouldn't they all give us the same answer? What is common to everyone's sensory experiences?

XARGOS.— I don't know. You tell me.

HERMES.— Isn't it common to everyone's sensory experience to observe a natural tendency for things to age, break down, or become disordered after the passage of sufficient time?

XARGOS.— If I've understood correctly, you're suggesting that sensory knowledge, acquired over time, leads everyone to accept that everything tends to break down naturally. If you want to convince me, you'll have to try harder. The way you've stated that, I don't believe it's true.

HERMES.— To me it's obvious...I don't understand why it's not to you.

XARGOS.— If it were true, what would be the most reasonable judgment we could make over any sensory thing? In other words, what would common sense say? To continue with your example, it would state that any sensory thing will break down. If we say that humankind has progressed in an orderly manner towards its future, we contradict what you just said. You would have to explain it, and even if what you said is true, can you find a clarifying example?

HERMES.— I think you're right, I'll try to find the error. Meanwhile, it's simple to find examples to verify it. I think anyone would agree that everything I'm about to say truly happens as I describe. Metal tends to rust, wood tends to fracture, machines tend to break down, living things sicken and eventually die, the roofs on houses tend to collapse, farmland is eventually exhausted, water flows downhill, cold things become warmer, hot things become cooler, glass tends to break, dirt tends to infiltrate homes, food tends to rot, and memory is lost to oblivion...

XARGOS.— Just a minute! Stop! I think that's enough. When you refer to so many kinds of disorder, I understand that you're leaving human intervention aside. Remember that humankind...

HERMES.— Of course, now the answer seems obvious to me. You were right to clarify this because common sense states that any sensory object, *left to its own devices; in other words, without intervention from any other being,* will end up breaking down if enough time elapses. I don't think you'll find anyone who believes otherwise.

XARGOS.— Of course, humankind had to intervene on the planet to achieve order. Humans were not isolated but had access to the planet's resources. Now, with this new premise, I don't think anyone will disagree. I certainly can't think of any example to contradict what you just said.

HERMES.— If there were a case to refute what we've said, it would probably be so strange that we would call it a miracle.

XARGOS.— Since it's impossible to find an everyday counterexample, let's accept what you've suggested as true.

HERMES.— Now we're getting close to something that all human beings truly share: the common sense we defined earlier includes a judgment common to all. Let's try to give it a more concrete form. We have the outer senses, each of which receives perceptions, or sensations. When these are combined, they generate sensory knowledge of an object, and this ability to combine the sensory information from all objects is what we call common sense. Because the senses do not provide the same information to all human beings, the knowledge obtained will differ from one person to another. But there is a characteristic shared by all objects regardless of the human being perceiving them. This quality is a natural tendency to break down, deteriorate, or become disordered...

XARGOS.— That's not true! I have another example of an object that doesn't become disordered... Living beings tend to order themselves.

HERMES.— Yes, they do, but they do so by disordering more than they order. They export disorder.

XARGOS.— I don't understand. Even though they disorder more, they are ordering themselves. So for them, that statement is not true.

HERMES.— In this sense, they are just like humans! It's true that living beings tend to order themselves and to organise. However, by arguing this, you do not deny the truthfulness of our statement. The key lies in what we said before: *without intervention from any other being*. Thus, animals, plants, and other living beings order themselves due to the intervention of other beings external to themselves. Therefore, common sense cannot make any statements about them. Their situation is no different from what we described for humankind. Don't forget that what we said is only true when they're left to their own devices; in other words, completely isolated.

XARGOS.— But...all living beings are related to each other. The fine china you dropped earlier broke precisely because it was not isolated; it smashed against the floor. If we had kept it isolated, it would never have broken. So, don't our theories apply to anything?

HERMES.— Yes, they do apply. We'll just have to be more careful when we discuss them. First of all, let's accept that if something is isolated, it either stays the same or becomes disordered. Second, let's accept that if something is not isolated, it either stays the same, becomes disordered, or orders itself. In our case, the china was not isolated: it hit the floor. The floor wasn't isolated because it received the impact from the china. However, both are virtually isolated from the rest of the Universe by the walls of the room and their remoteness. The rest of the world certainly had some influence, but we're sure it was small. If it had been significant, we would have included it to explain what was happening, and that wasn't necessary... After the impact, the floor remained largely the same, perhaps a little bit dirtier or chipped, and the china was broken. So, the final result for the floor and the china was disorder. In this case, all of the disorder went to the china. It was new before, but is now shattered all over the floor.

XARGOS.— I'm starting to understand. It's obvious that any object, over time, will either change or stay the same. So, you want to clarify that in our discussion the essential difference between objects lies in whether or not they are left to their own devices, without relating to anything else. In the first case, when a being is transformed, it always moves toward disorder. On the other hand, if it is altered by its relationship to another object, it might become ordered or disordered. It can never stay the same because we've said that it changed. However, in this last case, we know that if it ordered itself, the object it related to must have been disordered. In this case, we know absolutely nothing about what happened to the other object: it might have been ordered, disordered, or stayed the same. Is that right?

HERMES.— That's right. Look how likely an object is to move toward disorder. In isolation, it is either disordered or stays the same. If it's not isolated, it is disordered, ordered, or stays the same. So, the provisional score would be: Order 1, Disorder 2, No Change 2. We can see that order is starting to lose.

XARGOS.— I'm not very good with numbers, but I do understand that this method of scoring isn't very rigorous. In any case, it's obvious that an object is more likely to move toward disorder than order. With these numbers, the probability of disorder in an object known to have changed is two-thirds. The probability of order is only one-third. So, disorder wins when an object changes, but not by a mile.

HERMES.— That's right, you just summarised that natural tendency toward disorder very clearly.

XARGOS.— Thank you, even though it was you who guided me to this point. So, if we discover an object moving toward order, we know that it's not isolated and has therefore acted on something. But we also know that disorder has been created as a result of its action.

HERMES.— That's right. If we find something that's been ordered, we'll have to see what it disordered, and regard both as a single object. If we were to compare the sensory knowledge that several people have acquired over time concerning this set of objects, we would discover the natural tendency leading to deterioration or disorder. This discovery would be made by applying common sense, a person's ability to judge things reasonably. If we were to seek the existence of a reasoned judgment resulting from the common sense of all human beings, we would find precisely this. We could therefore describe the characteristic as *common common common sense.*

XARGOS.— Hey! What are you trying to do? Can't you just say it once?

HERMES.— It wouldn't be exactly the same because each of those three identical words refers to a concept that arose while we were speaking. The first one stands for the sense that combines the information received by all the other senses: that was the first definition we established. The second indicates that from this common sense, we only take the characteristic common to all objects when left to their own devices without relating to others: it refers to the deterioration and damage we spoke of. The third indicates that from this common sense, we take only the sensory knowledge common to all human beings: the judgment that can be made on the tendency toward disorder, which constituted the second definition we established. In other words, the *common common common sense* is the part of common sense common to all the senses, all objects, and all people. As you pointed out, *common sense* can be different from one person to another, but *common common common sense* will not be. If you agree, we can save ourselves the tongue twister by simply calling it common sense, even if we both know what we mean.

XARGOS.— Yes, we're referring to the common sense allowing all persons to reasonably judge that all sensory objects will tend to break down, deteriorate, become disordered, or some other unpleasant fate.

HERMES.— I think you just expressed it better than me.

XARGOS.— You do? I guess you're right because you took quite a long time to say what I just summarised in a single sentence. I was actually starting to get bored with such a vulgar, vulgar, vulgar word. And I'm still not convinced. That thing about living beings throws me off.

HERMES.— Before you could summarise it in a single sentence, we had to refine our ideas. You might understand it more clearly now, but it's not fair for you to discount our earlier efforts.

XARGOS.— The jokey tone was because you've convinced me: I accept that *common common common sense* says what it says universally. I just want to make sure we're not leaving any loose ends. I can think of plenty of beings that order themselves, but I can easily see that they're not all isolated. Let's assume the existence of the following counterexample: one being is ordered because it interacts with another that also orders itself, and that one can order itself because it interacts with the first. If such a case exists, what we said would affirm common sense would not. So, let's refine our ideas. I'd like to go back to the most complete example I can think of that could contradict what we're saying: living beings. I find this a delicate subject difficult to understand. Can two human beings mutually order each other by interacting, or can't they?

HERMES.— I find it difficult too, because what is happening to them is not simple. Both of us can only get a glimpse of the problem. Since we can't see the details, I'll try it a different way. If we accept that our common sense doesn't fail, we can say that living beings order themselves a little and create a greater disorder outside. In other words, the total overall result is disorder. If we add up everything they do, the balance is negative. If a man creates order thanks to another man, and the second man does so thanks to the first, there must be a third object, a third entity different from them, that is disordered to a greater extent than the order they've created. For example, all the food they consumed until they became adults. It's obvious that the food consumed has been destroyed. So, we can say that common sense does not fail because there is at least a third object disordered: the food. But you're right; it's not correct to simply generalise that all objects tend to be disordered. We saw this earlier, but you're right to stress how important it is to conclude that a being, if it changes, will only move toward disorder if it's completely isolated, without interrelationship to anything else. Common sense states that if something moves toward disorder, there's no need to worry, because this is normal. And if something is ordered, then something has surely broken down somewhere else, although it may be difficult to find. In your example, it could be the environment surrounding those two men who were ordered.

XARGOS.— In other words, for a being to become ordered, you have to consider its environment along with it in order to realise that overall, they will become disordered... What happens if we use the Universe as the environment?

HERMES.— The Universe is a very special object nothing like the beings we find in our macroscopic space. This is mainly because it has a property that our immediate environment will never have: according to its own definition, it is the only thing capable of containing *all* other objects. Our experience is based on what we feel in the limited space in which we are destined to live. I feel that extrapolating our experience, our common sense, to seek conclusions regarding the Universe would be very risky, and probably erroneous.

XARGOS.— Why can't we use our common sense to seek conclusions on the Universe?

HERMES.— I just told you. It has a property, a quality that defines it, that only it can have. That makes it very special.

XARGOS.— What is it?

HERMES.— The one I just told you about. It's the only object that contains all other beings. Our common sense has never experienced an object like that: it can only make assumptions about it. So, let's not go any further than we can. Let's apply our common sense only to what we know on our space-time scale. After all, that's where we're destined to live.

XARGOS.— That's a good idea, so let's put it to the test. Let's take a completely isolated room, where we've created a vacuum. If we place a nail inside it, nothing will happen. There will be no rust, disorder, or any events at all.

HERMES.— Common sense only states that if something happens, it will be an increase in disorder. Try introducing a guinea pig into that room, and you'll soon see that the life will leave its body due to the lack of oxygen: the most marvellous of orderings will have been lost.

XARGOS.— To summarise, when there are interrelationships between an object and its environment, common sense states nothing about the object: it can be ordered, disordered, or remain the same. When it is isolated, without intervention from any other being, it can be disordered or stay the same, but it can never be ordered.

HERMES.— That's right, as long as we've observed it for a sufficient time period; it's a matter of waiting long enough. And isn't this experience the same for any person, regardless of their religion, political ideas, nationality, or any other particularity?

XARGOS.— Yes, I think it is.

HERMES.— We're on the right track. I would like to stress that the sensory knowledge that human beings have on the tendency toward disorder of all sensory objects they experience is conceived by the common sense we've defined.

XARGOS.— In this common sense we're discussing, I think I see a similarity to what is known as the *Second Law of Thermodynamics.*

HERMES.— That's true to a large extent. And what greater support can our ideas have than something actually considered a law! But let's not be tempted to venture into science because it is beyond our capabilities right now. Let's leave that to minds that have the proper training, and keep our search as broad and general as possible. Let's not forget that we're trying to find a unifying link common to all humankind. So, it should not be framed by any particular profession, whether it be scientific, religious, or of any other stripe.

XARGOS.— You're right to clarify that. Let's leave the scientists, theologians, sociologists, and experts to their tasks, and see whether this common sense fits into all areas.

HERMES.— Let's hope that it does. If it doesn't, it would be difficult for it to serve as a unifying link.

3

On disorder and unhappiness – the natural order

Order is the placement of things in the corresponding place so that human beings have the greatest number of ways of doing things.

This definition is objective, global and universal.

Disorder implies unhappiness.

This statement is objective, global and universal.

XARGOS.— There's something I still don't understand clearly. In our discussion of common sense, we mentioned words such as deterioration, disorder, and ageing. Are they all synonymous, or could they all be encompassed by another word that would better define what we're looking for? I also noticed that all the words I just mentioned and the examples you gave earlier are loaded with a certain negativity. I think I can safely state that the vast majority of people would define the examples cited as undesirable phenomena, and agree that they cause unhappiness.

HERMES.— It's true that the word *disorder* came up.

XARGOS.— Came up? We've been using it all night to talk about common sense!

HERMES.— That shows just how much importance I believe we should attach to that word. Disorder refers to an alteration of the harmony inherent to something; in other words, a lack of order. And, depending on the context, it can easily be replaced by words such as confusion, discomposure, alteration, or excess. Order, on the other hand, can be defined as the placement of things where they belong. Order also refers to harmony or a good arrangement of things. Similarly, a rule or procedure observed for doing things is also a type of order. You were talking about the act of deteriorating, and I'm sure you'll agree that deterioration prevents an item from achieving the purpose for which it was intended. This purpose might be ornamental or operational, but in either case, it's easy to see how disorder between parts quickly causes an item to break down or deteriorate. Ageing translates to a loss of faculties in a thing or living being, and is similarly associated with a loss of order among the different elements of which it is composed. I don't consider these to be synonyms because disorder encompasses them all; a loss of order causes an item to break down or age. I've also noticed the negative connotation associated with disorder. So negative, in fact, that you mentioned unhappiness earlier. I would almost venture to suggest that disorder means unhappiness.

XARGOS.— I'm glad you also noticed the negativity implicit in the word disorder. However, I don't fully understand how we can view order in nature. As we said earlier, everything we experience in nature moves toward disorder. If one thing is ordered, another is disordered to a greater extent.

HERMES.— Let's use simple examples to try to clarify our concept of order in the world we live in. For instance, how do we know whether or not a deck of cards is ordered?

XARGOS.— For the deck to be ordered, the cards would have to be arranged by suit, and then maybe in ascending order.

HERMES.— If we chose cards randomly and I asked you how the deck was arranged, you would have a difficult time answering. What would your answer be if I had first sorted the deck, and then asked you about the position of each and every card?

XARGOS.— Knowing that it's sorted? I would say that I do know their position, but only if we had agreed on their arrangement.

HERMES.— You said yes because when the deck is sorted, there is no doubt in your mind about how the cards are arranged. This allows us to make blind decisions on which cards to take. But if the deck is unsorted, our intellect is lost when faced with a lack of information on which to base our decisions. Now let's suppose that I take the sorted deck and toss it into the air, allowing the laws of nature to act on the cards. When I pick up the deck, will it be ordered or disordered?

XARGOS.— Disordered. It's obvious that cards, clothes, or things scattered on the floor of a room are in a more disordered situation than when they're neatly arranged in a closet.

HERMES.— In the same way, we could play a game in which we place a huge cannonball on top of a mountain. If we went back after some time, wondering where the ball has gone, we could say that it's still on the mountaintop. We would have no doubt about its position. However, if by some chance the ball had succumbed to the passage of time and rolled down the side of the mountain, we would have trouble finding it because the number of places where it could be found would have increased drastically. In this sense, we could say that the ball is in a more disordered situation at the foot of the mountain than it was on the mountaintop.

XARGOS.— That's difficult to understand because I think it would be in a more ordered situation at the foot of the mountain. It definitely won't move once it's down there. However, if the ball is on the mountaintop, I'll always have some doubt about whether it's still there, or has already rolled to the bottom.

HERMES.— Obviously, the ball will have greater stability if it's at the foot of the mountain than on the mountaintop. But being on top is a more ordered situation than down on the valley floor. There's only one position at the top because that's the highest point. On the other hand, there are multiple possibilities on the bottom because that could be any of the points surrounding the mountain. Because we've said that nature always moves toward disorder, it's more logical for the ball to go from the mountaintop to the valley floor than vice-versa.

XARGOS.— On the other hand, a position at the foot of the mountain, perhaps marked by a unique colour in the sand, would have a similar order to the mountaintop because it would be the only point with that attribute.

HERMES.— But I'm sure you'll agree that when the ball falls, it will be difficult for it to roll naturally from the top to one position as opposed to any other. Otherwise, nature would not be tending to increase disorder, but to preserve it. According to what we discussed earlier, that goes against common sense.

XARGOS.— I think I understand what you're trying to say, but I don't share it. You're saying that the cannonball on the mountaintop is in a more ordered situation because nobody would have any doubt about its position; the top is the top. However, there are multiple possibilities at the foot of the mountain, and we have no way of knowing the exact position where it would fall. If we placed the ball on the mountaintop and wanted to go back for it, we'd only have to remember that it's at the top, and we'd be sure to find it. However, if we had placed it at the foot of the mountain, we would also have to note its exact position, since locating it visually would require additional effort. This relates order and information:

everyone knows that to order something means to put things where they belong, and this requires a prior agreement in order to specify such a place.

HERMES.— That's exactly what I'm saying. Just one clarification: placing a cannonball on a mountaintop would ensure that we'd find it, if it weren't for the fact that nature tends to seek disorder. If enough time passes, it's possible that the ball will roll down the mountain seeking a greater degree of disorder, thus breaking with the order we established. If I remember correctly, that was what bothered you earlier; knowing that the ball could fall. As I'm sure you'll agree, this is because you're well aware of the tendency, in common sense, toward disorder.

XARGOS.— I don't agree. You admitted that order requires the acceptance of a prior agreement, and agreements are arbitrary. They depend on the human beings who formulate them, and are therefore particular rather than universal. Consequently, the tendency toward disorder includes an arbitrarily defined concept that prevents it from being universal. Common common common sense cannot contain a definition different for each person! This definition of order is obviously not suitable for the purpose we are pursuing. If it works as you described, and there is more disorder as the number of possible locations increases, then why doesn't the ball, following the natural order, fly off into the immensity of outer space rather than remaining on this tiny planet called Earth, where we'll surely find it sooner or later? Or better yet, we can put a marble in a bowl. There's obviously one lowest point, at the bottom of the bowl, and a multitude of higher points. In this case, why doesn't the marble roll upwards by itself?

HERMES.— This time, you do seem to have taken me to a dead end. The way we were headed, we had stated that the only purpose of natural alteration is disorder, and this no longer seems to be the case. However, as everyone knows thanks to common sense, both the marble and the cannonball could eventually move to the lowest point. But once there, they would never roll upwards by themselves.

XARGOS.— Let's proceed cautiously. I think you're moving too fast. I don't know; one of the theses we've established is either not working, or is incomplete.

HERMES.— The thing is that nature, in addition to seeking disorder, facilitates downward movement and makes upward movement difficult, you know, because of the force of gravity. The effort required to move the ball downhill is very small; a slight push might be sufficient. But we'd have to make a large effort to push it back up once it stopped on the valley floor. This problem was solved by physicists nearly two hundred years ago, when they established the First and

Second Laws of Thermodynamics, leading them to define what they called *entropy*. As we can see in our dictionary, that's a measure of the non-usable portion of energy contained in a system. It's also a measure of the uncertainty in a set of messages, of which only one will be received. Conveniently enough for what we're discussing, it's even a measure of the disorder in a system: a mass of a substance with its molecules regularly ordered, forming a crystal, has much less entropy than the same substance in gas form, with its molecules floating free in full disorder.

XARGOS.— What you're doing now is physics!

HERMES.— The cohesion we seek must lie in the environment shared by all human beings, which is none other than the physical world. Physics is the science that studies the properties of matter and energy, considering only those attributes that can be measured. It always comes into play when such attributes are being discussed, and order is one of them. In any case, I haven't done any physics; that would be overstating it. I simply described the world that surrounds us, as a small child would do if called upon. I just used some technical terms at the end to clarify and further define some concepts.

XARGOS.— Now I do think we're going to get lost. We said we'd steer clear of any particular area of expertise, since we want humanity as a whole to understand us unanimously. Even though it may be impossible for all of humankind to understand in the same manner, we shouldn't use language that would only make us accessible to scientists or physicists.

HERMES.— I think you're right again. We should leave the correct definition of entropy to the physicists, and try to find that natural law from the perspective of two inexpert observers. Even so, don't forget that what we all share inevitably falls within an identical framework for everyone: common sense provided us with that experience.

XARGOS.— That should certainly be our position. But this time, allow me to protest vigorously. If I accept what you said earlier, about the most disordered situation being at the foot of the mountain, and about disorder meaning unhappiness, then being at the foot of the mountain would make people very unhappy. What nonsense!

HERMES.— You're right. What nonsense! As you said earlier, there's an error in the theses we established because otherwise we would all be astronauts drifting through outer space just like that cannonball! We need to reconsider this. I don't think the problem lies in our definition of common sense. Our mistake, as you rightly pointed out, was to oversimplify the definition of order. I think we can deduce that order is something more than a relationship that determines the location of the things we place in space. Going back to your example of the bowl and the marble, we can find more information about the definition of order. I'm sure you'll agree that there's a clear difference between the marble on the rim of the bowl and at the bottom.

XARGOS.— What difference?

HERMES.— When it's at the bottom, the marble won't start moving unless we bump it, or exert some additional force against it. But after a while, it will stop at the bottom again. From the rim, on the other hand, the marble will start moving by itself, with no effort on our part, as soon as we release it. As it falls, it will gain speed. It will roll over the lowest point, and then up to the opposite end of the bowl. It will stop there, roll back down in the opposite direction, and repeat this movement regularly. We all know it will happen this way, but do we know anything else?

XARGOS.— Yes. We know that every time the marble rolls down and then up again, it will never roll as high as the previous time. And this situation will continue until the marble eventually stops at the bottom of the bowl, which is the lowest point.

HERMES.— That's right. So, what has changed? Not the degree of spatial order of the marble, which has gone from the highest to the lowest point, both with a similar degree of spatial order. What has changed is the marble's ability to move by itself. The human being has also lost something: the possibility of performing some task with less effort. Thanks to its large mass, the cannonball on the mountaintop could be used to raise something to the summit with less effort, through the use of a pulley; provisions, for example. But at the foot of the mountain, the ball can no longer be used for this purpose unless we first make the effort to raise it to the top, in which case it would be better simply to carry up the provisions. The ball will never roll up to the top spontaneously, giving us the possibility of using it for our benefit. Remember when we said that order could mean placing things where they belong, harmony or a good arrangement of things with respect to each other, or the rule or manner observed for doing things?

XARGOS.— Yes, I remember.

HERMES.— And isn't using the cannonball to raise provisions with less effort a special way of doing things, and therefore a situation of order?

XARGOS.— Yes, it could be.

HERMES.— So, let's imagine that there are numerous high points, or different mountains, and only one low point, which is the valley floor. For our purposes, this situation is identical to the bowl. As we said in the beginning, if we only consider its spatial arrangement, when the ball falls from any cliff to the valley floor and stays there, it will be in a similar situation of spatial order. However, the possibility of raising provisions with less effort will have been lost. In this sense, we will have lost order, or gained disorder. If we accept once again that over the long term, all natural tendencies lead to disorder, we must conclude that the spatial order gained, which in this case will be null because it's gone from one specific point to another, will be less than the disorder generated. This is because we have lost the possibility of using the downward movement of the marble to perform some other task. We have less order because a method of doing things is no longer available to us.

XARGOS.— So, in order for what we said about common sense to be true, the key is to extend the definition of order as we have done. This is not complicated because the definition of order that we established right in the beginning already included all the meanings we needed; we just have to be careful not to forget them. So, we mustn't regard order as merely a spatial concept, only alluding to where things are placed, but also as something that says how things are done. Raising provisions to the top of a mountain using a cannonball is a way of doing things, and is therefore a situation of greater order than one without that option.

HERMES.— Moreover, losing the possibility of raising provisions to the top of the mountain with less effort might make us unhappy. That takes us back to something we hinted at earlier: disorder implies unhappiness.

XARGOS.— If that's the case, why would someone want to place a cannonball on a mountaintop, rather than a specific point at the foot of a mountain? Once on the valley floor, it would never move and produce that unpleasant situation characterised by loss of order.

HERMES.— You're very consistent in your approach to the question. Once again, you've shown that you feel uncomfortable with the idea of things changing position without your permission.

XARGOS.— I admit that it bothers me a bit. I'd almost say that I'd feel unhappy, depending on what things were to move. But you're not getting out of this. Answer me: Why would someone prefer the mountaintop to the valley floor as a place for the cannonball?

HERMES.— I'll take a risk, and generalise your question as follows: What is it that incites a person to act? If you express it that way, the approach is clearly pragmatic, and the question is not easy to answer. Let's suppose that we're concerned with where to place the ball because we're pursuing some purpose. In your view, what is the ultimate purpose that all human beings wish to achieve?

XARGOS.— From my perspective, I think I'd say happiness.

HERMES.— I agree. All of a person's actions would ultimately be geared towards seeking that state of mind that delights in the possession of a good, which we call happiness. Of the goods sought, we could highlight well-being, social recognition, love, comfort and, in general, the health of all beings. Similarly, we could define unhappiness as the misfortune or adverse circumstance that leads to the loss of a good.

XARGOS.— So, logically, happiness and unhappiness are opposite states, that cannot be present at the same time for the same good: each state is either present, or has been lost.

HERMES.—That's right, just as with order and disorder. Like I said, happiness is the state of mind that delights in the possession of a good. So, if that good is located where it delights the state of mind, producing happiness, that's because it's in the right place, where it will produce happiness. In this sense, we can say that it's ordered because order involves placing things where they belong. Happiness therefore implies order, and *non-order* implies *non-happiness*. When we sensed that disorder implied unhappiness, and thought it was merely an impression, we were on the right track: it is now a certainty. Indeed, disorder involves removing things from the places where they belong in order to make us happy, and will therefore make us unhappy.

XARGOS.— However, even though what you're saying is true, it's not useful to us. Remember that a definition of order based on specifying where each thing belongs, is particular. Each person will make rules to ensure that the resulting order will make them happy. Viewed in this way, unless we can solve this, the concept of happiness would not be universal.

HERMES.— You're right. That's how it happens in life, and that's how an inexhaustible source of conflict is maintained. Satisfying one's personal happiness, when based on rules that are also personal, strengthens one's sense of self, which differs among individuals. Unity is then made difficult if two different sets of rules fall on the same object. So, we should try to establish a universal definition of happiness. Let's say that happiness is the state of mind that delights in the possession of a good that enables a special way of doing things. If that good is found where it delights the state of mind because it provides a way of doing things, it produces happiness. It will therefore be placed where it belongs. We therefore accept that not all goods generate happiness, while also recovering the universality of our definition because both happiness and order essentially lie in the methods available for doing something, and these do not depend on a prior set of rules. Now, happiness does imply order.

XARGOS.— I find that quite striking, but obviously correct with the definitions you're using.

HERMES.— Note that according to the arguments we've used, nothing can be said about the proposition 'order implies happiness'. The only two certain propositions are 'happiness implies order' and 'disorder implies unhappiness'.

XARGOS.— That's too much logic for me. In any case, we obviously haven't said anything about the truthfulness of this statement. We shouldn't use it then, should we?

HERMES.— That's right. While all disorder leads to unhappiness, we don't know if all order leads to happiness. Consequently, certain orderings could lead to unhappiness. Remember that not all goods produce happiness, and a good is usually something ordered.

XARGOS.— So, we can't identify happiness with order or unhappiness with disorder.

HERMES.— No, we can't. We simply know what we already said: disorder causes unhappiness. So, we can only say that if we don't want to be unhappy, we should avoid disorder. This inevitably involves order. By avoiding disorder, we cannot be sure that we'll be happy, and we cannot be sure that we won't be happy. But by remaining in disorder, we will certainly not be happy.

XARGOS.— So, are you accepting that even by creating order, it's possible to be neither happy nor unhappy at the same time?

HERMES.— Yes.

XARGOS.— I agree. Let's call that state of mind *indifference*, if you agree.

HERMES.— I think that's a very good idea because it's a state of mind in which you don't feel inclined toward or disgusted by anything. It seems obvious that these three states exist: happiness, unhappiness, and indifference. We can use this uncertainty to verify that the statement 'order means happiness' is false. Suppose for a moment that it's true. Imagine that we're in disorder, and are therefore unhappy. Suppose that we remedy this by creating order where there was none and, since we've accepted that order means happiness, we would be happy. These assumptions indicate that one is either happy or unhappy, contradicting what we accepted earlier about the existence of indifference. We must therefore conclude that we have based our assumptions on a falsehood, which has led us to false conclusions. Order does not mean happiness because, at least in some cases, it can also mean indifference. Otherwise, indifference would not exist. Interesting.

XARGOS.— So, if there are orderings that produce indifference, and the boundary between indifference and unhappiness is very tenuous, perhaps there are orderings that produce unhappiness?

HERMES.— Obviously, there could be. For now, we know that fortunately all order does not lead to unhappiness. Another way of looking at this is to think that if all order meant unhappiness, there would be no happiness because all situations are either ordered or disordered. Because happiness exists, we must conclude that some types of order lead to happiness. On the other hand, because indifference also exists, we must conclude that other types would lead to indifference. The remaining ones, which you rightly said could also exist, would not be able to extricate us from the state of unhappiness.

XARGOS.— Calculating roughly, I find that happiness is quite a difficult state to achieve, and not only because the Universe tries to take it away through disorder. Even if we take the appropriate measures to change disorder into order, we only have one chance of success – happiness – compared to two chances of failure: indifference and unhappiness. Let's take the score (Order 1, Disorder 2) again and add the new information we're discussing. If order leads to happiness, unhappiness or indifference, depending on the situation, and disorder inevitably leads to unhappiness, we have a new provisional score: Happiness 1, Unhappiness 3, Indifference 1. Happiness loses by a mile: four points are scored against it for every goal it manages to score.

HERMES.— You're right, but that's how I think it happens in everyday life. That's probably why we have to do the right thing. That's why there are laws and rules which, with more or less effectiveness, seek to limit the areas of disorder and keep them inaccessible. That's also why we're reflecting on these things, and the reason there are professions, associations, and companies that seek, with more or less effectiveness, to bring order to our environment.

XARGOS.— It's easy to imagine examples that support the statement we know to be true. Can you think of any?

HERMES.— Wouldn't the cannonball on the mountaintop be helpful to our ascent when we were tired if we had previously connected a pulley to it? Of course it would! If it spontaneously rolled down, that would be associated with a lack of order in our method of climbing the mountain. Wouldn't a loss of health be associated with disorder in the body, or in the mind? Wouldn't the loss of those elements that meet our needs be subject to economic disorders or natural disasters? Does the natural order not inevitably lead to disorder, and therefore to the unhappiness of human beings?

XARGOS.— That does seem to be the case, but isn't it a contradiction to say that the natural order leads to disorder?

HERMES.— To my way of thinking, there is no contradiction here. The natural order is the way things are, exist and occur according to the laws of nature under which humankind lives. And, as we said earlier, common sense states that the laws of nature are characterised by a tendency toward disorder. I therefore reaffirm that the natural order leads to disorder, and therefore to the unhappiness of human beings.

4

On the possibility of the natural order being the unifying link sought

Human beings have the capacity to act on their environment.

This statement is objective, global and universal.

XARGOS.— So, humankind is doomed to be unhappy!

HERMES.— In a way. But because one wishes to be happy, one will strive at all costs to maintain the necessary order in oneself and one's surroundings; struggle to achieve the goods desired and not lose the ones attained. And yet, what is one person seeking to achieve order, when the enemy in this enterprise is none other than the Universe itself? Clearly an insignificant being. A difficult challenge; unless one is content with a small number of goods to be happy, failure seems probable. Setting one's sights higher, one will usually fail in the attempt. How to fight against disease, disasters, and accidents? An individual's powerlessness to achieve all of the goods necessary to remain happy is evident. These types of situation generate feelings of helplessness. Thus, a person's unhappiness springs from two sources: on the one hand, the natural order; and on the other, a powerlessness to contradict the natural order at all the levels one would wish.

XARGOS.— You're obviously referring to professional, family, and social lives, for example.

HERMES.— Yes, I'm referring to all the levels one needs to be happy.

XARGOS.— What a sad picture you're painting. A solitary human being, powerless to achieve everything he or she wants. Frustration will be present in each act, each attempt to struggle against the natural order will be thwarted. What choice does our human being have but to commend his or her actions to a higher power, and hope that fortune will smile down from above?

HERMES.— I think you're exaggerating because not all of his or her actions will fail. If our person is clever, only some will. Even so, I agree that it's a sad situation, but not tragic, because fortunately we've realised what is happening. It would be much worse to live this way unconsciously.

XARGOS.— I was only using pessimistic exaggeration to stress that the natural order produces disorder, and therefore unhappiness, nothing that you haven't said before.

HERMES.— My intention was, and is, to clearly express the nature of the sensory world around us. We can't do anything to change the natural order, which is why the common sense arising from it is indeed common, universal, and not at all arbitrary. But the fact that we can't modify the foundation of the Universe does not mean that we can't change our surroundings. If we don't like its fundamental character, that doesn't mean we should disguise it; we must simply take it into account. Once we've done that, we can change the world around us in the right direction, in order to strive for happiness. We just need to know how to do it without making mistakes.

XARGOS.— So this characteristic of the Universe, that we're calling the natural order and are aware of thanks to common sense, makes the world as harsh as it is.

HERMES.— Yes. However, the fact that it's so inhospitable could become a cause for joy.

XARGOS.— I don't understand. How can an adverse world become a cause for joy? What good does such unpleasant knowledge do humankind?

HERMES.— Obviously, we would all rather live in paradise, but the real world is neither paradise nor hell, but a Universe with a natural order as we've described. Once all humankind has accepted this, it will be easier to work towards a common objective. That is the answer to your question.

XARGOS.— So, how will humankind achieve that?

HERMES.— Humankind is the most complex link in the evolutionary chain. And in that complexity, humans acquired the most marvellous gift that distinguishes them from all others.

XARGOS.— I see, you're referring to intelligence. The typical stereotype about humans being intelligent. We see where that leads every day: to great technological advances and all manner of ill-treatment and humiliation.

HERMES.— There you go, exaggerating again!

XARGOS.— That's right, those are the two sides of the same coin. Reason employed at the service of humanity, and against humanity. This problem would be solved if humankind, in addition to intelligence, also had adequate morals.

HERMES.— If we do manage to conclude our work tonight, we will necessarily have discussed adequate ethics and morals. Your mistake, however, was to assume that I was only thinking of intelligence. You forgot another, more important gift.

XARGOS.— What can that be?

HERMES.— We've been talking about it for a long time. If a man wants to keep the unhappiness produced by the natural order out of his homeland, he has to strike back. But how can he do that, if he himself forms part of the natural order?

XARGOS.— I know where you're trying to go with this. You're asking me so I'll say it, and then you'll be able to refute it. But this time, I'm not going to contradict you or state it directly, so you'll have to say it yourself. Here's a clue, though: common sense enabled us to state that only when an object is completely isolated, the natural order, if it does anything, will disorder it. Fortunately, humankind is not isolated. Right?

HERMES.— That's right. The greatest gift we have is our ability to relate to our surroundings; in fact, that's why we exist. It enables us to outsmart the natural order by creating order wherever our second gift, intelligence, deems appropriate.

XARGOS.— Our intelligence wouldn't do us any good if we couldn't interact with our surroundings. But relating to the outside world wouldn't do us any good either if we didn't have the intelligence to do so as correctly as possible. For me, both are equally important. And I would also add others, such as memory, intuition, and curiosity.

HERMES.— Yes, but right now I don't know if what makes humankind different is reason, intelligence, memory, or any other faculty cleverly defined by psychology. Let's be prudent, and not make the mistake of turning our problem into a psychological issue. We're not experts on that subject, and we don't need to consider such nuances to reach our conclusion. Let's let psychologists discuss those matters, while we focus on the result of that distinction.

XARGOS.— And what can the result of that distinction be? What makes man different from all other beings? What can make him feel great? What can even enable him to unite with all other human beings, and forget the reasons for differentiation?

HERMES.— We already said that! The ability to create order on a small scale. Thanks to intelligence, to use the same word you mentioned earlier, human beings can manage to slow the progress of the natural order in some regions simply by interacting appropriately with the environment. An isolated human can only achieve limited results when trying to stop its progress. Each individual, depending on his or her limitations, will manage to fight it to a greater or lesser extent. But the best of humankind will never cease to be insignificant against the devastating progress of the natural order. Fortunately, this has turned us into gregarious beings, and we are not alone. We are a multitude of individuals struggling against the tendency toward disorder we have all experienced and included in our common sense. For humankind to advance, it is enough if each one of us gives our best effort to each small task we engage in.

XARGOS.— So, our enemies are not other human beings, or their political or religious ideas, or those of a different nationality or race. The true enemy that all of humankind faces, and which accepts no allies, is the natural order.

HERMES.— Yes, but with a difference. For now, this enemy is invincible, although humankind is gaining ground every day. If humanity has been able to advance without clear knowledge of its common objective, what would it not be capable of if every individual were to accept reality and the possibility that they could become a drop of water, water that would eventually turn the desert into a lush garden? I believe that this common sense, and humankind's ability to maintain order, is the true link that could keep all humanity united.

XARGOS.— It's interesting that you would phrase it that way because that very concept contains the germ of its own destruction.

HERMES.— I don't understand. What do you mean?

XARGOS.— It seems like there would be different criteria about how to act to achieve order. The situation would inevitably lead to different ways of thinking that would soon become differentiated and break the unity formed.

5

On the existence of a universal categorical imperative

A comparison of the efforts associated with the do and undo actions, performed with maximum efficiency, establishes a moral criterion.

This criterion is objective, global and universal.

HERMES.— I think you're right. We need to investigate whether there's a single way to proceed in this search for order, so there can be no difference of opinion. How should we start?

XARGOS.— Let's unquestioningly follow Christian doctrine. See how simple that is?

HERMES.— We said no religions. I wonder how we could convince Islamic scholars of that?

XARGOS.— You're right, we agreed that the unifying link would be based on the experience of common sense so as to be universal. That way, individuals could seek righteousness in their actions based only on the information provided by their common sense and their logic.

HERMES.— Have you noticed that all actions taken to achieve order are associated with some kind of effort? Perhaps the effort is larger depending on the magnitude of the order to be achieved. Let's try looking at it that way.

XARGOS.— I don't quite understand. Effort and order are related? Could you give me an example?

HERMES.— I'll try. Isn't knowledge a good that involves ordering the mind for those who acquire it?

XARGOS.— I think so.

HERMES.— And isn't knowledge acquired through study?

XARGOS.— Yes.

HERMES.— Doesn't study involve a large effort for those engaged in it?

XARGOS.— Yes.

HERMES.— And doesn't knowledge introduce order into the mind, order that did not previously exist?

XARGOS.— Yes. You already said that.

HERMES.— I think knowledge is a good example of a desired good that requires effort to be achieved.

XARGOS.— Now I need to interrupt our dialogue to see whether I understand the idea of effort the same way. For me, effort can be one of three things: the energetic use of physical force against an impulse or inertia; the energetic use of one's vigour or will to achieve something by overcoming difficulties; or the use of costly elements to pursue a purpose. Does it mean the same to you?

HERMES.— Yes. Again, dictionary definitions are beneficial for clarifying our ideas. And the moral of the story is: Never go into the wilderness to engage in philosophical reflection without mosquito repellent and a good dictionary.

XARGOS.— Actually, I'd rather use yours! Let's see what you think of this. Imagine a student who's trying to decide whether or not to study. He knows studying requires effort, and he rebels against that. If he didn't have a tutor to guide him and inform him of the benefits of studying, how could he approach this decision?

HERMES.— Having reached this point, I think this question is easy to answer. If an indecisive person has consciously absorbed the information provided by his common sense, and knows that achieving happiness depends on the order surrounding him, he will not hesitate to choose the option that generates the greatest amount of order. Since we've seen that creating order involves effort, that choice should be the one that requires the greatest effort. In the student's case, it would be to study.

XARGOS.— If I've understood correctly, you're suggesting that the greatest effort can be the criterion for finding the path that leads to improvement.

HERMES.— Yes, but let's proceed carefully. I'm only examining the possibilities so we can find the solution by a process of elimination. The criterion of maximum effort could quickly lead to unreachable or impossible means or solutions, which would exceed the abilities of the human being wanting to attempt them.

XARGOS.— Not only that, but if more order means more effort, we would find ourselves unhappy due to overwork. The very path to happiness would work against our happiness. I don't think that's a good criterion.

HERMES.—That's true. Even without taking it so far, there could be situations where several decisions predictably lead to the same result, but nonetheless involve various degrees of effort. If we used the criterion described above, we would always choose the solution requiring the most effort. We would waste energy by performing excessive work. So, I think human beings should seek the solution requiring the least effort, of all the options available for achieving the desired order. In other words, humans should be effective, while also remaining efficient.

XARGOS.— When you say human beings should be effective, do you mean they should successfully meet their objective of achieving the optimal level of order? And when you say they should be efficient, do you mean they should achieve that order with the least possible effort?

HERMES.— That's right. We all know that to achieve happiness, we must obtain goods, which can be tangible or intangible. And, as a necessary part of reaching that happiness, we all wish to work as little as possible.

XARGOS.— I'm still finding this discussion a bit paradoxical. On the one hand, we're saying that the more order we wish to achieve, the more effort we should devote. This has led us to suggest that given two possible courses of action, we should always choose the one requiring the greatest effort. But now, once we've chosen the solution that produces the most order, we have to seek other possible solutions that also generate the same amount of order. Of all those, we have to choose the one requiring the least amount of effort. Not only does this seem like a contradiction, but couldn't the minimum amount of effort be even less than the effort required to perform one of the actions we had originally ruled out?

HERMES.— The answer is no, since we maintained that the solution we ruled out generated less order than the one we accepted. Otherwise, this last statement would not be true.

XARGOS.— I'm not sure I understand. Let me approach this another way. I don't know which action is the right one: *a, walking,* or *b, running,* but I do know that *running* involves more effort than *walking,* so I decide that option *b* must be the right one. I've been effective, but now I'm going to be efficient. I look for other actions that could produce the same effects as action *b*. Of all those, I choose the one requiring the least amount of effort. I'll call this *c, taking the bus.* Now I compare the effort required for option *c* and option *a*. I find that *a* requires more effort than *c*, so I should have chosen *a* instead of *c*. What's the problem? Isn't that inconsistent?

HERMES.— Obviously, the answer to the same question formulated with letters is also no. I got a bit confused with the letters, but I understood you. The problem you described is not an inconsistency; you just haven't finished solving it. You were right to choose action *a*, which means you were effective at that moment. However, you could still be inefficient because you haven't sought the other actions that produce the same effects as *a* and require less effort. If you do that, you'll be back at solution *c*. And this case you described is not the most interesting one. It would be more productive to compare two different types of actions: cleaning the house and watching television, which I'll call *a* and *b*. Let's suppose that by doing this, we find an action that replaces *a, cleaning the house,* while being associated with the least of all efforts. We'll call this *d, hoovering.* A greater effort may be required for *d* than for *c, watching television while lying on the couch, remote in hand.* In that case, we would choose *d*. If it did not require more effort, we would stick with *c*. Either *c* or *d* would be the effective and efficient solution. In this case, hoovering obviously requires more effort, so the solution would be *c*. This person would have concluded that he should clean before watching TV. Since you were the one who started naming things with letters, I'm sure you haven't had any trouble following me.

XARGOS.— Actually, I did. Sorry, but I still don't understand. This recipe for deciding when to act is not ideal because it can lead to undesirable situations that clearly do not produce order. Let's look at a practical example with an obvious solution. Imagine a builder who doesn't want his tools to get ruined outdoors. If he gets his wish, that will make him happier, or at least it won't make him unhappy. But he would like to do nothing for months. That will make him happy, or at least he thinks it will. To avoid making a mistake, he decides to reason by seeking order.

HERMES.— Right, let's consider that. The builder could decide to leave all of his tools completely unprotected, or he could build a shed and store them there. Not doing anything is associated with zero effort, and apparently much happiness. Building a shed will force him to work, which will apparently make him unhappy. But that's not true. Of these two possibilities, he should lean towards building the shed because this option is associated with more effort. In this case, it is also associated with greater order and therefore the option of achieving happiness. Once he has chosen the option of building the shed, he could start a new discussion about whether or not to use tools. The option of building a shed without any tools obviously involves more effort than building with tools. However, the order achieved in both cases would be practically the same. A shed would exist where there wasn't one before. In this situation, where the order achieved would be the same, he should choose the option requiring the least effort, which would obviously be to build using tools.

XARGOS.— Let's see if I understand. Imagine that someone else is out walking, and encounters the roof of the shed, built with the first man's effort. He feels an irresistible urge to tear this construction down and reduce it to rubble. He thinks this will make him happy, although he's not quite sure. But because he's a person, like us, who has discussed order and effort, he decides to first analyse whether his actions are right and will lead to the greatest happiness for all concerned. He considers the possibility of not doing anything, which involves zero effort, and the possibility of demolishing the roof, which would undoubtedly require some effort. Because he is well aware that achieving order is associated with effort, he concludes that destroying the roof would definitely result in happiness for him, and decides to tear it down. He also decides to destroy it using the tools stored inside, which will require less effort.

HERMES.— I see that you're being very shrewd. That's good because it'll help us see how erroneous our thinking was when we directly associated effort with order, and therefore with the possibility of keeping unhappiness at bay. Just to clarify, this character was wrong to think that order will cause him happiness. That only holds true if you make a good enough choice. Not all types of order lead to happiness, remember? On the other hand, you were right to doubt the rule we suggested earlier. As you so brilliantly proved with this example, effort does not necessarily imply order. Tearing down the roof requires effort, and involves imposing disorder where order had been present.

XARGOS.— How should we proceed, then?

HERMES.— I think we should resume our discussion at the point where we said that achieving order is associated with effort. Now, thanks to your observation, we've seen that disordering also requires effort. It's obvious that effort is not enough to distinguish between order and disorder. However, now that I've considered it more carefully, the path we've chosen may not be so slippery. We said that human beings should be efficient as well as effective, which means they should seek minimum effort. Wouldn't you agree, Xargos, that the effort required to achieve disorder will be less than the effort needed to restore the order destroyed?

XARGOS.— That's exactly what I meant when I said that your reasoning seemed erroneous and tried to clarify it, without much success, using letters. However, I couldn't prove it because that problem was somewhat different from this one, since actions *a* and *b* had no specific objective. As soon as the objectives became clear, when we discussed the builder and the man who wanted to destroy the roof, it was obvious that our theses were not entirely correct. That makes it impossible to identify order simply by seeking an associated effort. I find it paradoxical that we have this difference of degree that you pointed out, but we can't use effort to identify order as our intuition suggests. How can there be an imbalance between the efforts associated with order and disorder, as you're suggesting? Is this imbalance real, or is it a product of this philosophical mess we're making?

HERMES.— It's real. We already established that the natural order tends to disorder all sensory objects. Thus, when we use our efforts to produce disorder, we'll always have the natural order as our ally, which will make its own contribution to our efforts. On the other hand, when we want to create order, the natural order will be our enemy, forcing us to make a greater effort. We'll have to make the effort needed to achieve order, in addition to the effort required to defeat the natural order. This is the key for differentiating between order and disorder, a key we didn't have before. But you had already begun to sense it, when you formulated the question with letters.

XARGOS.— That's right. So, we can't use actions requiring the least amount of effort as our criterion, since that would lead us directly to disorder and therefore unhappiness. Neither the greatest nor the least amount of effort are valid criteria as we had considered them. However, I find it peculiar that order and disorder are differentiated as you said before. Is there a way forward?

HERMES.— I think there's still a way forward. It involves comparing *doing* with *undoing*. Let me explain. Before taking any action to achieve a particular good, or a certain amount of order, we have to analyse all possible means of achieving it. Of those possibilities, we have to choose the most efficient, or the one requiring the least amount of effort. That means *doing*. Before proceeding to take that action, we have to imagine what would happen if we later wanted to put everything back the way it was before we took the action. To meet this new objective, we evaluate each and every action we can think of to achieve our purpose, and choose the most efficient. That would mean *undoing*. We then compare the effort required to take the action, and the effort required to undo it. Of these two options, we choose the one representing the greatest amount of effort. That action is certain to help order our immediate surroundings, and will therefore take us closer to happiness.

XARGOS.— This does seem like a reasonable path for distinguishing between what we should and should not do. The builder, while considering whether or not to build a shed, would first imagine all possible ways of doing it, and would choose the one requiring the least amount of effort. He would then imagine the shed already built, and each and every imaginable way of eliminating it, and would choose the one requiring the least amount of effort. Finally, he would compare the effort required to build the shed and the effort required to destroy it, and would discover that it's costlier to build it than it is to destroy it. This conclusion would tell him that to build the shed creates order, while destroying it creates disorder, so he would undoubtedly choose to build it. The second character, who encountered the roof by chance, would first imagine all possible ways of building it, and then choose the least costly. He would then imagine the roof after it was destroyed, and try to think of all possible ways of rebuilding it before eventually choosing the one requiring the least effort. He would compare the effort required to build the roof and the effort required to destroy it, and discover that the former is greater than the latter. By choosing to tear it down, he would clearly be creating more disorder, and therefore less happiness for humankind. He would therefore decide not to destroy the roof, and he would be doing the right thing. Even if it made him momentarily happier, the poor builder would probably feel much unhappier after finding his shed reduced to rubble. The balance would clearly be a victory for the greatest enemy of all: the natural order.

HERMES.— I can see that you understood my reasoning quite well, and you've helped me consolidate my ideas, which were still pretty shaky. To see if we really agree, shall we apply this way of thinking to the student who was wondering whether or not to study?

XARGOS.— I'm happy to give you a chance.

HERMES.— In this case, the student is debating whether to laze around or to study. He decides that studying will provide him with knowledge, but doesn't know if the effort is really worth it. He decides to consider all possible ways of acquiring this knowledge, and thinks of teaching himself, learning with a private tutor, or going to school and studying. Of these possibilities, the one that provides knowledge with the least amount of effort is attending classes at school and studying. His objective is to order his mind through knowledge. In order to be efficient, he's determined the least costly way of achieving that. He now considers the effort required to undo that order in his mind and lose the knowledge acquired, and concludes that the simplest and least costly way to do that is by skipping class, not using the acquired knowledge at all, and letting time pass in this way while oblivion takes its toll. The effort required to achieve this is virtually zero. So, given these two possibilities, and leaving his immediate urges aside, he decides to study. He knows that in the long run, this helps to create order where there wasn't any before, and therefore takes him closer to happiness.

6

On the importance of objectives and dialogue in the imperative

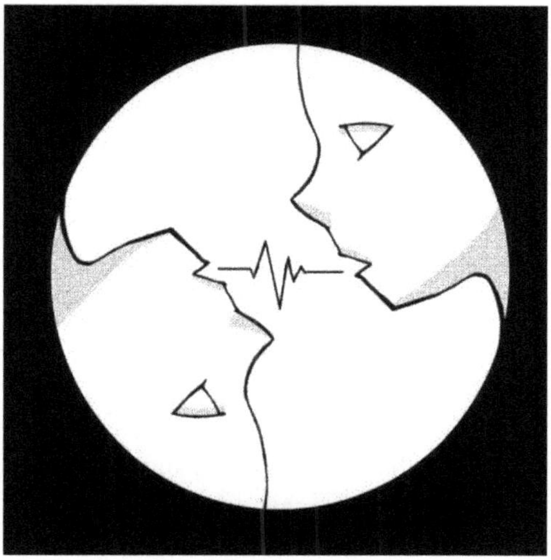

Dialogue sets the objectives.

The objectives define the do and undo actions.

These statements are objective, global and universal.

XARGOS.— I think we agree because I would have expressed the solution similarly, but I still have a question about the criterion we've been discussing. When we started using letters, I sensed that an objective is essential to the formulation of our criterion.

HERMES.— I don't understand. Do you mean that when we consider taking a particular action, we should first establish an objective or purpose we want to achieve with that action?

XARGOS.— That's exactly what I mean.

HERMES.— I hadn't considered it carefully. Why don't we solve this by using our criterion with and without an objective, and see where those two paths lead us?

XARGOS.— I think that would be a good idea. Let's consider an everyday situation. A man is tired of shaving, and considers whether or not he should continue doing it.

HERMES.— Why don't you apply this action on a larger scale?

XARGOS.— OK. Someone is tired of removing hair from their legs, and considers whether or not to continue doing so. They apply our criterion, and choose the hair removal method that requires the least amount of effort. They imagine the action, and then consider the least costly way of undoing it. They realise that it's impossible to restore exactly the same hairs to their legs. Observe, Hermes, that what I've described as impossible is restoring exactly the same hairs that are now in the trash. Given these two situations, our criterion would recommend the action requiring the greatest effort, which would be to return those same hairs to the person's legs, if that were possible. That person would obviously conclude that they should not wax their legs.

HERMES.— So, if they haven't considered any objective, they conclude that it's better not to remove the hair from their legs?

XARGOS.— Yes, but they could have considered the question differently. Immediately after wondering whether or not to remove the hair from their legs, they could ask themselves another question: What is the purpose of waxing my legs? The person would answer that their objective would be to feel more attractive. But let's imagine that they want to further examine their answer by applying our decision-making criterion. In my view, they should proceed as follows: My objective is to feel as attractive as possible. In order to do so, I must wax my legs, which is the least costly way of removing leg hair. I now imagine my legs after they are waxed, and seek the opposite objective, which is to feel less attractive by having hairy legs. To achieve that, the least costly method is to wait for the hair to grow back, which requires zero effort. Because they must choose the costliest solution, they decide to wax their legs, and are happy because there is agreement between reason and their wishes.

HERMES.— This time, the action did have a purpose, and they decided to wax their legs. This shows that you were on the right track when you said that having an objective is essential to the formula we're investigating.

XARGOS.— Yes, the objective is important, but this example still holds some mystery. This person could have taken things even further, and reasoned as follows: My objective is to feel attractive all the time, not just for a while. I'm therefore thinking that a necessary preliminary objective is to never have hairy legs at all. To achieve that purpose, the person therefore considers actions such as waxing their legs every week for the rest of their days, or laser hair removal treatments to solve the problem once and for all. Their experience tells them that laser hair removal would be the solution requiring the least amount of effort overall. Once they've chosen the action involving the least effort, they would imagine it completed and establish the opposite objective, which would be having hairy legs at least once. They would imagine actions that could lead to this, and find that there are none. The hair would have been removed permanently, and would never grow back. That would therefore be an action beyond the capabilities of a human being, which would mean that they should not choose the laser hair removal option. This would create an emotional conflict for this person, whose desire is to never have hairy legs again. Fortunately, this conflict does not exist because people who want to feel attractive know why they're doing it, and do not use such reasoning when choosing hair removal methods. But what should we do, as people who are using this example to find the scope of our decision-making criterion? On this occasion, there was also an objective, but we reached a different conclusion than in the previous case.

HERMES.— It amazes me that an example so uninteresting for uniting humankind, which you'll remember was our objective, can create such serious concerns. Obviously, if we trip up while applying our criterion to such apparently simple questions, what will happen when we try to decide on each and every one of the issues that concern the spirits of human beings throughout their lives? We therefore have no choice but to try and solve these matters.

XARGOS.— Yes, but I still really don't know what's wrong. A categorical criterion should not produce contradictory answers. However, the questions we formulated seemed very valid to me. What are we forgetting?

HERMES.— First, let's analyse why we reached different answers when considering the two cases with an objective. What's actually happening is that the person's desire is bound to destroy something that was previously ordered by a being who watches over that order day and night. That leg hair grew on the person's skin thanks to the effort expended by their body. The natural order would be happy to destroy that hair, just like the person who wishes to remove it. If you don't believe this, my receding hairline can provide living proof. So, the situation is complicated because the human body, among the unconscious functions it undertakes each day, is in charge of making hair grow back. The solution to this problem seems complicated because when we take actions to destroy that hair, we are adding our conscious efforts to those of the natural order, which also tends to destroy, and both forces are working against the human body's effort to restore lost order by making the hair grow back. Depending on the means we employ, we may or may not defeat the body's efforts. And if we do defeat them, we may do so temporarily or permanently.

XARGOS.— The contradiction does seem to stem from the interference between our acts of will and the code recorded in our genes. We already discussed that a while ago, but it still surprises me. The body is an example of something that orders itself, so doesn't that go against the natural order we established?

HERMES.— You're assuming from the beginning that our intellect can act to create order. However, although the evidence is clear, you're reluctant to believe that the unconscious and instinctive part of ourselves can achieve this. If we have time to dig deeper, we'll try to reflect on this later. Please remind me. But for now, let's not lose sight of our hair removal problem. We now have to assume that our intelligence is not the only thing capable of identifying and creating order, and that there are other beings also capable of doing so, including the unconscious part of ourselves. That means that our decision-making criterion will have to assess whether our purpose, or objective, is contrary to that of another person. In that case, we would look like two ants pulling on the same bread crumb from opposite directions, expending a large effort but achieving nothing. It's obvious that if humankind were united with a unanimous action criterion, we would manage to avoid those situations. For now, thanks to your instructive example, we've been warned that there can be decision-making situations in which our interests are mixed with those of another being. If two beings are acting in opposite directions, with one undoing what the other does, each one will realise that it takes more and more effort to perform the act. They will therefore conclude that they have all the more reason to do so – their mistake lies in confusing the natural order's tendency to disorder with the activities of another being.

XARGOS.— How fascinating! We suggest a problem to prove the importance of an objective, and consequently discover that we have to be careful with purposes involving other individuals. What should be done?

HERMES.— In this case, I don't think they should act until they agree on their objectives. A joint decision should be made by means of dialogue. The dialogic process would define a common objective and common actions, making it possible to jointly apply the criterion to decide on how to act. Needless to say, the situation would become more complicated if rather than two individuals with intermingled interests, there were three, four, or even a multitude.

XARGOS.— It seems to me that our criterion wouldn't be worth much in that situation. I'm sleepy, and I don't think we're getting anywhere. Wasn't the national team playing today?

HERMES.— No one said the path would be easy. If anything, we've spoken from the beginning of the difficulties we were likely to encounter. So let's not be surprised now. If it was anyone's fault, it was whoever established the natural order the way they did, if they were actually able to choose. We were only left with the possibility of struggling against it, which is already enough, and should be taken as a challenge regardless of who or what made it that way. Some would view it as a divine test, and others as a natural cause. What do we care about causes, when we're all forced to live immersed in the Universe just as it is? Let's understand it and adapt to it, while seeking our happiness in an orderly fashion.

XARGOS.— I see that you're truly committed to the problem. And since the nearest bar with a TV is far away, I won't leave you alone right now. I think the last thing you said is true. Also, the natural order is the same for all human beings, and because it seems adverse to our happiness, it must be the link we were seeking for humankind. The human motto should be: *All Against the Natural Order! Let's not make its path any easier.*

HERMES.— Listening to you, it seems that this way of thinking strengthens you, or maybe you've already got your mind on the playing field. If it makes you stronger, as I hope it does, that proves that the path we're on is not far from achieving our goal of universal union. That maxim you just coined could be a good start.

XARGOS.— I'm not out on the playing field. I'm more intrigued and committed than ever, much more than when you caught me with your plan for the night. And I must admit that I think it's because we're on the right track.

HERMES.— To avoid overloading our weak brains, and prevent dialogue problems as we had with our leg hair discussion, shall we continue reasoning as if we were the only beings in the Universe capable of generating order in the struggle against the natural order?

XARGOS.— I don't think we should throw in the towel so easily. We should force ourselves to successfully conclude the work we started. If there's going to be a reason for all of humankind to unite, I don't understand why we should now proceed with such a monstrous limitation as assuming that all of humanity consists of a single person. And dialogue is precisely what arose naturally during our inquiries. If it's so important that it could change the result of our conclusions, I see no reason it should be neglected.

HERMES.— Now I'm the one who feels the need to stop, rest, and organise my thoughts, but I'm going to continue.

XARGOS.— Good choice. That tiredness is probably caused by the order we're creating in our little heads, where before there was none. That means we're fighting against the natural order, which is obviously exhausting.

HERMES.— I'll take your advice and keep humanity multitudinous. However, this distinction is not so important because the dialogue between all of the individuals comprising humankind should ensure a single common objective. From that moment on, we can regard humanity as a single whole.

XARGOS.— That's true, and it was the goal we established earlier, to get humankind to behave as a single whole. We seem to be finding our way naturally.

HERMES.— In our discussion on leg hair removal, we reduced humankind to mind and body, and encountered an important clash. When the objective was to *feel attractive*, depending on the action performed, the conclusion could be either to remove the leg hair or not. We saw that the interests of the body and mind were in direct conflict: the objective imposed by the person's rational mind went against their own body, which proved to be another being with the ability to create order. To avoid this contradiction, our criterion must include the need to identify beings capable of creating order, and establish a dialogue with them if their interests and ours meet along the way.

HERMES.— Why don't you increase the scale of the action again?

XARGOS.— What do you mean?

HERMES.— We went from shaving to hair removal on a large area of the body. Why don't we go from a person's skin to the planet's surface, and from body hair to the flora and fauna that spontaneously appear on the Earth? What prevents a person from deciding to destroy a wooded area? What motivates them to do it?

XARGOS.— They might decide to do away with the plants, insects, and animals that live there in order to build a road, neighbourhood, or factory. It's absurd to consider a dialogue with the insects in the region! Of course! That's just as absurd as a dialogue with leg hair!

HERMES.— You obviously can't have a dialogue with leg hair. When I talk about dialogue with an unconscious being, I mean establishing an internal discourse of questions and answers to characterise the raison d'être and the peculiarities of that being. The objectives are obviously in the people wanting to perform the action, not in the unconscious being. The ethical framework lies in the people wanting to perform the action. The measure of the order lost or gained lies in the human being. As we said, more or less order exists depending on whether there is a larger or smaller number of special ways in which things can be done. A factory provides a special way of doing things that was not present before, so a factory increases order, but also transforms the surrounding area. If an insect becomes extinct after its habitat is destroyed, the active ingredient for an undiscovered medicine might disappear along with it, thus eliminating numerous special ways of doing things. At the very least, it would eliminate the special ways of doing things of all the people saved from a certain death by that medicine. So, if we can't assess how many special ways of doing things are lost and how many are gained, it's difficult for us to know whether the situation results in increased or decreased order. Again, dialogue appears a very useful element for resolving this conflict. And to ensure a correct assessment, the dialogue should include as many people as possible. On such a path, it's plausible to think that the large number of people involved could result in an enormous number of objectives. For example, there could be people seeking new active ingredients, new relationships, new possibilities, or new ideas regarding existing organisms. In particular, there could be someone whose objective is to study the insect and its habitat before extinction. The factory builder would then have to engage in a dialogue with that person before construction, until common objectives could be found. In general, avoiding the extinction of life forms is just such an objective.

XARGOS.— I think you're starting to get side-tracked. Considering everything we've said so far, do you feel capable of establishing a method for guiding our actions?

HERMES.— I think so. Let's establish an action criterion as follows: First, we'll establish an objective clearly being pursued. If an action is considered directly, we'll have to try to identify an objective based on the action. Second, we'll try to establish whether our objective has interests that are intermingled, for better or worse, with the interests of other beings that are endowed, consciously or unconsciously, with the ability to act. If such beings are present, we'll try to start a dialogue with them in order to establish a common objective satisfactory to everyone. Third, once an objective has been established that is shared by all parties, we'll analyse it, and come up with a list of all possible actions that could help us achieve the objective as it was formulated. Fourth, we'll choose the least costly action in order to achieve the objective. Fifth, we'll establish an objective the exact opposite of our initial goal. Sixth, we'll establish once again whether this objective interferes with anyone else's objectives and, if so, determine whether the goal opposite to our original objective is common to all parties. Seventh, we'll list all actions that would help us meet this new objective in order to select the least costly one. Eighth, we'll compare the two actions chosen, one of which is associated with the initial objective and the other with the opposite goal, and choose the least costly. This way, we would know which action to choose; in other words, we should know how to act.

XARGOS.— Yes, this is quite a rational way to proceed, now that we've seen the problems that can arise with the leg hair scenario. But let's not forget that the question we were trying to answer was whether or not an objective was needed. With the hair removal scenario, we also saw how the result achieved using our method changed when we considered the problem with or without an objective.

7

On whether one should act for the sake of acting

Acting for the sake of acting generates disorder.

Not acting generates disorder.

These statements are objective, global and universal.

HERMES.— I'm aware of the distinction between the problem stated with and without an objective, and that's why I included the objective clearly in that last description of our method. I don't think the action criterion works if no objective is specified. When applied without an objective, the same conclusion is always obtained. In the leg hair scenario: not to wax; and for actions in general: not to act. I find this conclusion so interesting that I'd like us to analyse it, if you don't mind.

XARGOS.— I'd be delighted! I'm always the one who's eager to reach some kind of conclusion.

HERMES.— Let's try to reason generally. We're considering whether or not to take a particular action when we don't have a specific objective beyond the action itself. This action will require a certain amount of effort, which we are capable of assessing, and we conclude that the action is feasible and reasonable. We then imagine the action already completed, and try to establish the opposite actions necessary to restore the object we've acted upon to its original state. But after thinking for a while, we'd discover that there's no possible action that can be performed.

XARGOS.— What do you mean?

HERMES.— The action, when performed, will generate order or disorder, depending on the action. The opposite action chosen will obviously have to restore the initial order or disorder, as appropriate. In addition, the contribution made by the natural order, which will always tend to disorder, will accompany our activities in either of the two human actions considered.

XARGOS.— That's true, but there's a difference between the natural order and us: it always disorders globally, while we order and disorder locally. In other words, it has access to all objects, while we only have access to a few. Right?

HERMES.— That's right. The action, whatever it is, acts upon an object that undergoes an alteration. But for the object to be affected by our action, we ourselves must be affected by the object's reaction. We're both immersed in the natural order established in this Universe. If something is modified, it must be with an overall tendency toward disorder. This way, the set formed by us, the object, and most likely the surrounding environment will be disordered when the action is performed. And when we perform the opposite action, whatever that may be, the same set will tend to become disordered again. The set will be disordered twice in a row. The conclusion is that it's impossible to perform an action exactly opposite to the original. There will always be something that's left differently from the way it was: it might be us, the object, the environment, or all of the above. In any case, someone or something must tend to receive the disorder generated. We can conclude that the exact opposite action is impossible, and constitutes an unfeasible effort. According to our criterion, we must choose it. In other words, we must choose not to perform the action.

XARGOS.— I find this conclusion surprising. We've learned that if we don't know the raison d'être for an action, or if that action has no purpose, the action should never be performed. What better criterion for not doing anything when we don't know for what purpose we'd be doing it! I think everything we're saying is very reasonable because our decision-making criterion just told us to be prudent, and not waste our effort doing things if we don't know why we're doing them. I think that's fantastic. Don't do anything if you don't know why you're doing it! It sounds like a proverb. Not only that, but I would say that it sounds like *common sense*.

HERMES.— Right. If you agree, let's consolidate our ideas, because we've learned some important things.

XARGOS.— The way you're speaking is quite doctrinal. It almost sounds like a sermon.

3

The Participation of the European Union and its Member States in the Food and Agriculture Organization of the United Nations

DONATA RUGARABAMU[1]

I. Introduction

The relationship between the Food and Agriculture Organization of the United Nations (FAO or the Organization) and European Union (EU) is long-standing, reflecting positive cooperation in pursuit of the shared goal of eradicating poverty, hunger and malnutrition. The year 2021 marked the thirtieth anniversary of this successful collaboration.

On 26 November 1991, at the Twenty-sixth Session of the FAO Conference, the FAO Member Nations voted to admit the European Economic Community (EC) as a Member Organization of the FAO. The admission of the EC required amendments to the FAO Constitution and other elements of the Organization's *Basic Texts*. It took more than two and a half years of arduous negotiations concerning the form that this membership would take to establish this new legal framework.

The EC became the 161st Member of the Organization and it is still, at the time of writing, the only Member Organization of FAO. This was also the first time

[1] This contribution and the research behind it were possible thanks to the contribution of Ana María Fernández De Soto.

that a regional economic integration organisation was accepted into the membership of an agency of the United Nations system. Currently, FAO's membership is composed of 194 Member Nations, one Member Organization, and two Associate Members. It has a presence in over 130 countries worldwide.

This contribution describes FAO's mandate, the history of the negotiations between FAO and the EC that led to the creation of a *sui generis* form of membership based on the concepts of 'alternative exercise of rights' and 'transfer of competence', the scope and attributes of membership of Member Organizations in FAO, the complexities that have arisen in the exercise thereof, the strategic role played by the EU in FAO, and the partnership between FAO and the EU.

II. FAO's Mandate

FAO was created on the understanding that the only way to tackle global challenges pertaining to food and agriculture, and to ensure that people have regular access to enough high-quality food to lead active and healthy lives, was through collaboration amongst nations.

Following the 1943 United Nations Conference on Food and Agriculture, which committed to the establishment of a permanent organisation focused on food and agricultural issues, an Interim Commission for Food and Agriculture was created to prepare the draft Constitution of FAO. This constituent legal instrument was formally adopted at the First Session of the FAO Conference held in Quebec, Canada, in 1945 by 34 Nations. On 16 October 1945, FAO was officially established to lead the efforts to defeat hunger and improve nutrition and food security globally. The international 'World Food Day' is celebrated annually on 16 October – the anniversary of FAO's foundation – to promote global awareness and action to tackle global hunger and malnutrition.[2]

The objectives of the Organization are listed in the Preamble of the FAO Constitution as:

> raising levels of nutrition and standards of living of the peoples under their respective jurisdictions; securing improvements in the efficiency of the production and

[2] United Nations General Assembly, Resolution A/RES/35/70, adopted at its 83rd plenary meeting, 5 December 1980.

distribution of all food and agricultural products; bettering the condition of rural populations; and thus contributing towards an expanding world economy and ensuring humanity's freedom from hunger.[3]

As a specialised agency of the United Nations,[4] FAO is an autonomous organisation with its own governance structure. It was brought into relationship with the United Nations through the conclusion of a relationship agreement in accordance with Articles 57 and 63 of the United Nations Charter.[5]

The legal framework of FAO is set out in its *Basic Texts* which include, inter alia, the FAO Constitution, the General Rules of the Organization (GRO), and the Financial Regulations adopted by the FAO Conference pursuant to Article IV.2 of the Constitution.

The Conference is the primary Governing Body of the Organization. It is attended by all Members and Associate Members of FAO, together with Observers from non-Member Nations and from inter-governmental and non-governmental organisations. Each Member Nation has one vote. The Conference meets for regular sessions once every two years. Extraordinary sessions may also be convened at the request of Members or if the Council so instructs the Director-General.[6]

The Council is the executive organ of the Conference and acts on behalf of the Conference between its sessions. It normally holds at least five sessions in a biennium. It is chaired by an Independent Chairperson of the Council, who is elected by the Conference and may serve a maximum of two two-year terms of office. The Council is composed of 49 Members, elected by the Conference and distributed among regional groups.[7]

There are three Council Committees that report to, and assist, the Council in the performance of its functions. The Council Committees are the Programme Committee, the Finance Committee and the Committee on Constitutional and

[3] FAO Constitution, *Basic Texts of the Food and Agriculture Organization, Vols I and II*, 2017 edn, Preamble, https://openknowledge.fao.org/server/api/core/bitstreams/437ae43e-7426-4e48-9d99-623dde284258/content.

[4] Article XII of the FAO Constitution provides that 'The Organization shall maintain relations with the United Nations as a specialized agency within the meaning of Article 57 of the Charter of the United Nations'.

[5] See Agreement between the United Nations and the Food and Agriculture Organization of the United Nations, available at https://treaties.un.org/Pages/showDetails.aspx?objid=0800000280166076&clang=_en. It entered into force on 14 December 1946 by the approval of the General Assembly of the United Nations, which followed approval by the FAO Conference.

[6] FAO Constitution, Article III.

[7] FAO Constitution, Article V and General Rules of the Organization, *Basic Texts of the Food and Agriculture Organization, Vols I and II*, 2017 edn, Rules XXII, XXIII, XXIV and XXV, https://openknowledge.fao.org/server/api/core/bitstreams/437ae43e-7426-4e48-9d99-623dde284258/content.

Legal Matters (CCLM).[8] They are bodies of restricted membership, comprising appropriately qualified representatives of Member Nations elected by the Council.[9]

The Council is further assisted by four technical committees covering the sectors under the mandate of FAO: the Committee on Agriculture, the Committee on Commodity Problems, the Committee on Fisheries, and the Committee on Forestry.[10] Membership is open to all FAO Members. They report to the Council on programmatic and budgetary matters pertaining to their area of work, and to the Conference on global regulatory and policy matters.[11]

The Governing Bodies also include Regional Conferences for Africa, Asia and the Pacific, Europe, Latin America and the Caribbean, and the Near East. They normally meet once every biennium in non-Conference years. The Regional Conferences provide a forum for consultation on all matters pertaining to the mandate of the Organization within a region, including any special issues of interest to the Members of the region concerned.[12] (An Informal North America Regional Conference was established in 2010).

Since its establishment, FAO has seen many changes in its governance structure. Over the years, the *Basic Texts*, including the Constitution, have been amended to improve the Organization's efficiency and effectiveness within a changing global system to ensure that the Organization remains well positioned to deliver its core mandate.[13] The admission of the EU to membership of FAO is one such example of change and adaptation in order to remain fit for purpose.

III. The Creation of a New Legal Framework in FAO

A. Request for Admission by the EC

The FAO Constitution adopted in 1945 permitted only nations to become Members. This was consistent with the fact that, at that time, they were the only entities recognised in international law as having the capacity to undertake legal

[8] FAO Constitution, Article V.6(a).
[9] General Rules of the Organization, Rules XXVI, XXVII and XXXIV.
[10] FAO Constitution, Article V.6(b).
[11] General Rules of the Organization, Rules XXIX, XXX, XXXI and XXXII.
[12] General Rules of the Organization, Rule XXXV.
[13] In addition, while not an FAO Governing Body, the Committee on World Food Security (CFS) is an intergovernmental committee established under the FAO Constitution. The CFS reports to the Conference and to the United Nations General Assembly (UNGA), through the Economic and Social Council (ECOSOC) and the Conference. It is open to all Members of the Organization, the World Food

commitments in relation to matters under the purview of the Organization's mandate. However, when states transferred their legal competence to enter into international binding commitments to a regional organisation, it became necessary to revisit the governance architecture of the Organization to respond to this new reality and make better use of multilateral channels.

This is precisely what occurred in FAO: in a letter received by the Director-General on 17 May 1989, the Foreign Minister of Spain, who was at the time the President of the Council of Ministers of the EC 'recalled [that] the present observer status of the EEC ... had become inadequate in view of the extensive transfer of competence from the Member States to the EEC in areas of FAO's operations'. Accordingly,

> he requested that the matter be placed before the FAO Council so that the Director-General might be given a mandate to hold discussions with the EEC Commission with a view to devising a status of member that would be commensurate with the present competence of the EEC.[14]

At the time of this request, the EC had observer status with FAO. Representatives from the EC could attend the plenary meetings of the Conference, the meetings of any commission, of any technical committee of a commission and of any technical committee. They also had the right to speak in meetings of the Governing Bodies and, upon request to the Chairperson, to participate in the discussions. While, normally, observers could only take the floor after all Member Nations had made their statements, a practice emerged whereby, pursuant to a request by an EC Member State, the Chairperson would give the EC the floor to speak on behalf of its Member States, sometimes before all FAO Member Nations had spoken.

Nevertheless, the EC, as an observer, had no right to vote, nor could it participate in the work of the drafting committees of the Governing Bodies. The EC was also not eligible to become a party to treaties and agreements – concluded pursuant to Article XIV of the FAO Constitution – setting up, inter alia, regional fisheries bodies.

Programme and the International Fund for Agricultural Development, or non-Member States of the Organization that are Members of the United Nations or any of its specialised agencies. See General Rules of the Organization, Rule XXXIII.

[14] Committee on Constitutional and Legal Matters, Report by the Secretariat to the CCLM regarding the possible forms of Membership for Regional Economic Integration Organizations in FAO including possible amendments to the Basic Texts, Fifty-sixth Session, Rome, 15–17 April 1991 (CCLM 56/4), paragraph 1. The Communication from the Government of Spain Regarding the Status of the EEC with Respect to FAO is reproduced as Appendix K to the Report of the Ninety-fifth Session of the Council, Rome, 19–30 June 1989 (CL 95/REP), https://www.fao.org/governing-bodies/council/reports/en.

Considering the transfer of competencies that had taken place under the EC legal framework, neither the EC nor its Member States felt that the constitutional arrangements in FAO enabled sufficient participation by the EC in the work of FAO.

Expanding the scope of FAO membership to regional economic integration organisations (REIOs) was a complex matter that needed careful consideration: the transfer of competence from Member States to the EC applied only in some aspects of food and agriculture. Accordingly, since competence on many key areas still lay with EC Member States, any form of membership in FAO had to be flexible enough to reflect the respective competences. Furthermore, any new form of membership could not give additional participatory or decision-making authority within the FAO Governing Bodies, but had to ensure that the exercise of rights would alternate between the EC and its Member States, as appropriate. Finding the suitable balance was challenging.

B. History of Negotiations and Guidance from the Council

After lengthy discussions, the Council, at its Ninety-fifth Session in June 1989, invited the Director-General

> to explore the options for a form of membership of FAO of regional economic integration organizations to which their Member States had transferred competence in some fields of activity of FAO, along with the full constitutional, legal, financial, and other implications for the Organization of such options.[15]

At the Ninety-eighth Session of the Council in November 1990, the issue of REIO membership was discussed in more depth. The Council identified a number of basic considerations with respect to possible membership of REIOs to be taken into account in the course of further talks with the EC and the preparation of further documentation and possible draft amendments to the *Basic Texts*, namely:

1. Any membership status for regional economic integration organizations in FAO would have to be of a *sui generis* nature. Moreover, any solution adopted would

[15] Report of the Ninety-fifth Session of the FAO Council, Rome, 19–30 June 1989, paragraph 276 (CL 95/REP), https://www.fao.org/4/t0175e/T0175e06.htm.

have to be applicable to any regional economic integration organization to which competence had been transferred in fields of activity of FAO.

2. The membership status would have to reflect the alternative nature of the exercise of membership rights. This implies that either the regional organization or its Member States, but never both, would exercise these rights, depending on which had competence in a given matter. The sum of the rights and obligations of a regional organization and its members should never exceed the total rights and obligations of the Member States thereof.
3. The respective competence of a regional economic integration organization and of its Member States would have to be defined in a precise and complete manner. In the case of the EEC, it appears that at present, broadly speaking, competence in matters relating to the institutional life of the Organization remains with the Member States; thus, the ECC would, at present not be eligible to be a member of restricted bodies such as the CCLM, the Programme Committee or the Finance Committee.
4. An amendment to Article XIV of the Constitution in order to allow participation by regional economic integration organizations in agreements concluded thereunder should be considered as a matter of urgency.[16]

For over two and a half years, discussions were held with the EC and deliberations took place within the Council's subsidiary bodies to carefully design a new institutional legal framework allowing membership of FAO by REIOs. Several drafting proposals to effect amendments to the Organization's *Basic Texts* were presented and considered by Members. A Committee consisting of a limited number of Member Nations selected by the Director-General, after consultation with the regional groups, was set up to review the texts and, in particular, those provisions on which consensus had not been reached.[17]

The Council was always mindful that, if REIOs were to become eligible for membership in FAO, this would create an important precedent for other organisations of the United Nations system.[18]

[16] Committee on Constitutional and Legal Matters, Report by the Secretariat to the CCLM regarding the possible forms of Membership for Regional Economic Integration Organizations in FAO including possible amendments to the Basic Texts, Fifty-sixth Session, Rome, 15–17 April 1991 (CCLM 56/4), paragraph 6. See also Report of the Ninety-eighth Session of the Council, Rome, 19–30 November 1990, paragraphs 233 to 242 (CL 98/REP), https://www.fao.org/4/t0424e/T0424e06.htm#6.

[17] The Committee of Member Nations Established to Review the Proposed Amendments to the Basic Texts of the Organization to Allow for Membership of Regional Economic Integration Organizations in FAO (REIOC) held its First Session from 9 to 13 September 1991 and its Second Session from 2 to 4 October 1991 under the Chairmanship of Ambassador Bula Hoyos (Colombia). Its Reports were transmitted to the CCLM.

[18] Committee on Constitutional and Legal Matters, Report by the Secretariat to the CCLM regarding the possible forms of Membership for Regional Economic Integration Organizations in FAO including possible amendments to the Basic Texts, Fifty-sixth Session, Rome, 15–17 April 1991, paragraph 11, (CCLM 56/4).

Following this hectic period of negotiations, the Council, at its Hundredth Session held in November 1991, unanimously agreed to transmit to the Conference for its consideration a compromise text containing proposed amendments to the *Basic Texts* enabling membership of REIOs.[19]

IV. Amendments to *Basic Texts* of the Organization and EU Membership of FAO

At its Twenty-sixth Session, on 18 November 1991, the FAO Conference adopted Resolution 7/91 amending the *Basic Texts* to allow for a unique form of membership in FAO by REIOs.[20] This new membership is distinct from that of Member Nations both with respect to its nature and to the terms under which it is exercised. The Resolution entailed amendments to Articles II, XIV and XVIII of the FAO Constitution and introduced a new Part into the General Rules of the Organization.[21] In accordance with Article XX.2 of the FAO Constitution, the amendments entered into force immediately. The following are the most important elements of this unique form of membership.

A. Eligibility Criteria

The definition of the term 'regional economic integration organisations' and the criteria that such organisations should meet to become FAO Members was fundamental, considering that the Council had requested that any proposal should be applicable, on an equal basis, to any REIO.

To be eligible for membership, REIOs must meet the criteria established under Article II.4 of the Constitution, namely:

(i) the REIO must 'be one constituted by sovereign States, a majority of which are Member Nations of the Organization';

[19] Report of the Hundredth Session of the FAO Council, Rome, 5–8 November 1991, paragraphs 117 to 125 (CL100/REP), https://www.fao.org/4/t0545e/T0545E05.htm#5.

[20] Report of the Twenty-sixth Session of the Conference, Rome, 9–27 November 1991, paragraph 302.

[21] See Part D of the General Rules of the Organization concerning 'Member Organizations' (Rules XLI to XLVI). Amendments were also introduced to Rule XIX of the General Rules of the Organization.

(ii) the REIO must be an organisation 'to which its Member States have transferred competence over a range of matters within the purview of the Organization'; and

(iii) this transfer is to include 'the authority to make decisions binding on its Member States in respect of those matters'.

Article II.3 of the Constitution provides that, for those REIOs meeting these criteria, the Conference may, by a two-thirds majority,

> decide to admit as a Member of the Organization any regional economic integration organization ... which has submitted an application for membership and a declaration made in a formal instrument that it will accept the obligations of the Constitution as in force at the time of admission.

At the time of adoption of these amendments to the FAO Constitution, there were many emerging REIOs. However, the eligibility criteria, in particular the requirements to have a transfer of competence and binding decision-making power in the areas of FAO's mandate, were not met by many of those bodies. Indeed, these restrictive criteria have limited the opportunities for membership by other REIOs of FAO, with the EU being the only Member Organization to date.[22]

B. Transfer of Competence

The FAO Constitution provides, under Article II.5, that

> [e]ach regional economic integration organization applying for membership in the Organization shall, at the time of such application, submit a declaration of competence specifying the matters in respect of which competence has been transferred to it by its Member States.

The Constitution further clarifies, under Article II.6, that 'Member States of a Member Organization shall be presumed to retain competence over all matters in respect of which transfers of competence have not been specifically declared or notified to the Organization'.

[22] It is recalled that as of the entry into force of the Treaty of Lisbon on 1 December 2009, the EU replaced and succeeded the EC. Accordingly, as of that date, the Delegation of the EC accredited to FAO has become the Delegation of the EU.

The Constitution also stipulates under Article II.7 that

> [a]ny change regarding the distribution of competence between the Member Organization and its Member States shall be notified by the Member Organization or its Member States to the Director-General, who shall circulate such information to the other Member Nations of the Organization.

For FAO membership purposes, the transfer of competence must include 'the transfer of treaty-making powers by Member States and means that complete power with respect to that subject is transferred and that no residual power remains with the Member States'.[23]

Further provisions concerning competence of Member Organizations in FAO are included in the GRO, with a view to clarifying the concept and avoiding potential differences in interpretation. For example, pursuant to Rule XLII.1 of the GRO, Member Nations may request a Member Organization or its Member States to provide information as to which, as between them, has the competence with respect to specific matters under the purview of FAO. The GRO make it imperative that the Member Organization or its Member States provide such information upon request.

The GRO further stipulate, under Rule XLII.2, that, prior to any FAO meeting, the Member Organization or its Member States shall indicate which, as between them, 'has competence in respect of any specific question to be considered in the meeting and which, as between the Member Organization and its Member States, shall exercise the right to vote in respect of each particular agenda item'.

In the case of the EU, the declaration of competence made at the time of its application for FAO membership was made in broad and general terms.[24] It stresses that the scope of the competence which the Member States have transferred to the EC is, by its nature, subject to continuous change and that the EC will make further declarations whenever the need arises. In practice, prior to every FAO Governing Body session in which it intends to participate, the EU transmits statements of competence and voting rights for each agenda item. These statements are

[23] See Commentary to amended paragraph 4 of Article II of the FAO Constitution in Resolution 7/91, contained in paragraph 302 of the Report of the Twenty-sixth Session of the Conference, Rome, 9–27 November 1991, https://www.fao.org/4/x5587e/x5587e0b.htm#vii.%20constitutional%20and%20administrative%20matters.

[24] The declaration of competence made by the EC is enclosed with the Instrument of Acceptance of the FAO Constitution submitted by the President of the Council of the European Communities on 25 November 1991.

circulated as official documents, and their contents are announced at the beginning of each session.[25] It is observed that the actual allocation of competence is entirely an internal matter for the EU and its Member States; it is not a matter in respect of which FAO takes any decision.

C. Alternative Exercise of Rights

The basic principle under which Member Organizations may exercise their membership rights is enshrined in Article II.8 of the FAO Constitution. The principle states that '[a] Member Organization shall exercise membership rights on an alternative basis with its Member States that are Member Nations of the Organization in the areas of their respective competences and in accordance with rules set down by the Conference'.

In simple terms, this provision means that membership rights can only be exercised by either the Member Organization or its Members, never by both at the same time.

V. Scope of Rights and Obligations of FAO Member Organizations

The rights and obligations of Member Organizations are fundamentally different from those of Member Nations. The *Basic Texts* as amended by the FAO Conference in 1991 set out the scope and limitations under which REIOs may exercise their membership rights.

The amended *Basic Texts* did not extend Member Organizations' rights to include all those rights enjoyed by Member Nations.

[25] See, for example: Forty-second Session of the Conference, 14–18 June 2021, 'Statement of Competence and Voting Rights Submitted by the European Union and its Member States' (C-2021/INF/2); Hundred and Seventy-first Session of the Council, Rome, 5–9 December 2022, 'Statement of Competence and Voting Rights Submitted by the European Union and its Member States' (CL 171/INF/3); Twenty-eighth Session of the Committee on Agriculture, 18–22 July 2022 'Statement of Competence and Voting Rights Submitted by the European Union and its Member States' (COAG/2022/INF/3 Rev.1); Thirty-fifth Session of the Committee on Fisheries, 5–9 September 2022, 'Statement of Competence and Voting Rights Submitted by the European Union and its Member States' (COFI/2022/INF/17).

A. Voting Rights

Article II.10 of the Constitution provides that

> a Member Organization may exercise on matters within its competence, in any meeting of the Organization in which it is entitled to participate, a number of votes equal to the number of its Member States which are entitled to vote in such meeting. Whenever a Member Organization exercises its right to vote, its Member States shall not exercise theirs, and conversely.

In observance of the general principle of the alternative exercise of membership rights, if the EU votes on an issue, then its Member States do not do so, and vice versa. Accordingly, the EU's total voting weight within FAO does not increase as a consequence of the EU's membership.

Agendas of the Governing Body sessions normally cover a variety of issues. When agenda items touch upon both matters in respect of which competence has been transferred to the Member Organization and matters which lie within the competence of its Member States, the exercise of membership rights may be more complex. However, the General Rules of the Organization clarify, under Rule XLII.3, that both 'the Member Organization and its Member States may participate in the discussions' but 'the meeting, in arriving at its decisions, shall take into account only the intervention of the party which has the right to vote'.

The declaration of competence submitted by the EU at the beginning of each session is, therefore, critical to guide a chairperson's management of a session.

B. Participation in FAO Governing Body Sessions

In accordance with Article II.9 of the Constitution, except as otherwise provided

> a Member Organization shall have the right to participate in matters within its competence in any meeting of the Organization, including any meeting of the Council or other body, other than bodies of restricted membership …, in which any of its Member States are entitled to participate.

It is essential to distinguish between the right to be present and the right to participate. Following consideration by the CCLM and the Council, it was concluded that the EC could be present in Council sessions but, on matters where exclusive competence remained with its Member States, the EC could not participate.[26] A review of the past practice of the Council shows that the *Basic Texts* have been interpreted as entitling the EU, on matters where competence is shared between the Member Organization and its Member States, to participate in the deliberations on the report of the Council, through the delegation holding the rotating Presidency of the Council of the EU.

The FAO bodies of restricted membership, where Member Organizations cannot participate, are: 'the Programme Committee, the Finance Committee and the Committee on Constitutional and Legal Matters'.[27] In addition, the General Rules of the Organization also bar participation of Member Organizations in the following bodies: 'the Credentials Committee or the General Committee or any other body of the Conference dealing with the internal working of the Conference as the Conference may decide'.[28]

Moreover, Member Organizations shall not hold office in the Conference and Council or any of its subsidiary bodies.[29]

C. Participation in Other FAO Governing Bodies and the Committee on World Food Security

The EU, as a Member Organization, also participates in meetings of other FAO Governing Bodies (and their subsidiary bodies), such as the Committee on Fisheries (COFI), the Committee on Agriculture (COAG), the Committee on Commodity Problems (CCP), and the Committee on Forestry (COFO). The EU also participates in and contributes to the discussions of the Committee on World Food Security (CFS) and the sessions of the FAO Regional Conference for Europe. It is recalled that it participates in accordance with the statement of competence that it submits at the beginning of each session.

Thus, for example, while it may not be involved in decision-making in some areas under FAO's mandate, through its participation in technical committees, the EU has been involved in treaty-making in areas in which it has competence, such as fisheries. For example, the EU contributed to the technical consultations for

[26] Report of the Hundred and Third Session of the Council, Rome, 14–25 June 1993, paragraphs 237 to 242 (CL 103/REP). Report of the Sixtieth Session of the Committee on Constitutional and Legal Matters, Rome, 14–16 April 1993.

[27] General Rules of the Organization, Rule XLVI.

[28] General Rules of the Organization, Rule XLIII.

[29] General Rules of the Organization, Rule XLIII.3 and Rule XLIV.

the drafting, and in 2009 voted for the adoption by the COFI, of the Agreement on Port State Measures to Prevent, Deter and Eliminate Illegal, Unreported and Unregulated Fishing.

D. Participation in FAO Statutory Bodies

Statutory Bodies are established in accordance with Articles VI and XIV of the FAO Constitution to address any technical subject falling within FAO's mandate, from aquaculture to livestock, or from controlling desert locusts to forestry-related issues. There are more than 100 Statutory Bodies operating in FAO. They include standard-setting and regulatory bodies.

While participation in these bodies may appear to touch upon aspects related to the institutional life of the Organization, agreements concluded under Article XIV of the FAO Constitution – international treaties – have a separate legal existence of their own. Accordingly, the issue of membership rights of Member Organizations in these bodies will be based upon the competence and the provisions of the instruments themselves.

For instance, the EU has been a member of the General Fisheries Commission for the Mediterranean (GFCM)[30] since 27 July 2000 and, similarly, a member of the Indian Ocean Tuna Commission (IOTC)[31] since 27 October 1995, as their constituent agreements allow for this participation.

The EU also adhered to the International Plant Protection Convention (IPPC) on 6 October 2005. However, as the subject-matter governed by the IPPC is a shared competence, the exercise of rights by the EU in this forum tends to be more limited than in fisheries bodies, an area where the EU has exclusive competence.

Participation of Member Organizations in the CODEX Alimentarius Commission was amply discussed when amendments to the Organization's *Basic Texts* were under discussion. CODEX is a joint FAO–World Health

[30] The Agreement for the Establishment of a General Fisheries Council for the Mediterranean (GFCM) was approved by the FAO Conference at its Fifth Session in November 1949. In accordance with Article IX, the Agreement came into force on 20 February 1952, the date of receipt of the fifth instrument of acceptance. The Agreement was registered with the Secretariat of the United Nations on 5 April 1952 under No 1691.

[31] The Agreement for the Establishment of the Indian Ocean Tuna Commission was approved by the FAO Council at its Hundred and Fifth Session, 25 November 1993, under Article XIV of the FAO Constitution, by Resolution No 1/105. As provided for in Article XVIII, the Agreement entered into force on 27 March 1996, the date of deposit of the tenth instrument of acceptance. The Agreement was registered with the Secretariat of the United Nations on 3 June 1996 under No 32888.

Organization (WHO) statutory body established by the Eleventh Session of the FAO Conference in 1961, entrusted with the establishment of international food standards, guidelines and codes of practice to contribute to the safety, quality and fairness of international food trade, which are a reference for the World Trade Organization.[32]

The FAO Council, at its Ninety-ninth Session held in June 1991, addressed the question of a Member Organization's participation in CODEX. A legal opinion from the FAO's Legal Counsel, with the concurrence of the UN Legal Counsel, concluded that membership in FAO would entitle a Member Organization to participate in bodies operating jointly with other organisations in the same way as in the work of the Organization, subject to the necessary amendment of the statutes and rules of procedure of the bodies concerned.[33] Due to the relevance of the matters discussed therein, and the potential impact on trade, the EU and its Member States actively participate in CODEX sessions.

E. Other Limitations on Membership Rights of REIOs

The General Rules of the Organization set out some additional limitations on the membership rights of Member Organizations. For example, 'Member Organizations shall not participate in voting for elective places as defined in paragraph 9(a) of Rule XII'.[34] Thus, for matters such as the election of the Independent Chairperson of the Council and the Director-General, only Member Nations may vote. As the General Rules of the Organization may apply *mutatis mutandis* to other bodies, the restriction may also apply to such other bodies.

Another matter in which Member Organizations may not participate pursuant to Article XVIII.6 of the Constitution, relates to the approval of and contribution to the Regular Programme Budget of the Organization. This is consistent with the fact that, rather than paying an assessed contribution to the Regular Programme Budget, each Member Organization pays 'a sum to be determined by the Conference to cover administrative and other expenses arising out of its membership in the Organization'.[35] Normally, the sum due from the EU is paid

[32] Report of the Eleventh Session of the Conference, Rome, 4 to 24 November 1961, Appendix D.
[33] For the text of the legal opinion, see Report of the Ninety-ninth Session of the Council, Rome, 10–21 June 1991, Appendix I (CL 99/REP), https://www.fao.org/4/t0482e/T0482E16.htm#16.
[34] General Rules of the Organization, Rule XLV.2.
[35] FAO Constitution, Article XVIII.6.

into a trust or special fund established by the Director-General under Financial Regulation 6.7.[36]

VI. Practical Matters Arising in the Exercise of Membership Rights by the EU

Over the years, the operationalisation of the amendments introduced to the Organization's *Basic Texts* to allow membership by REIOs have required numerous deliberations by the FAO Governing and Statutory Bodies to assess what these amendments have meant in practical terms. Discussions and guidance provided by Members have contributed clarifying aspects of the scope of this membership. Below are some examples of practical matters that have arisen in the exercise of membership rights by the EU and the guidance that has been provided for their resolution.

A. Seating Arrangements

In the course of the Hundred and Second Session of the Council, held in November 1992, a practical question arose as to the seating of the EC within the Council chamber. While a provisional seating arrangement had been agreed upon by all Members at the suggestion of the Independent Chairperson of the Council, further clarity and a long-term solution was required. Accordingly, the Council requested the CCLM to examine the subject and submit a definitive solution for its consideration in June 1993.[37]

After a review of the applicable legal framework, the CCLM concluded that there was

> a distinct difference between the legal basis on which Member Nations on the one hand, and Member Organizations on the other hand, may be entitled to participate in meetings of the Council. Member Nations may participate if they are elected to

[36] Forty-second Session of the Conference, 2021, 'Payment by the European Union to Cover Administrative and other Expenses Arising out of its Membership in the Organization' (C 2021/LIM/7).

[37] Report of the Hundred and Third Session of the Council, Rome, 14–25 June 1993, paragraph 237 (CL 103/REP).

> membership of the Council under the provisions of Article V.1 of the Constitution. Member Organizations, on the other hand, were expressly declared ineligible for election to any body of restricted membership by the terms of Article [II.9]. Their rights to participate in Council meetings derived then, not from Article V.1 of the Constitution, but rather from Article II.9, which accorded Member Organizations 'the right to participate in matters within its competence in any meeting of the Organization, including any meeting of the Council …, in which any of its Member States are entitled to participate'.

The CCLM also noted that

> [i]t appeared that this difference in legal status was well reflected in the provisional seating arrangements presently in force, which clearly reflected the sui generis nature of Member Organizations' rights to be present, participate and exercise membership rights in Council meetings.[38]

The Council 'adopted the proposal by the CCLM that in future sessions the EEC should continue to be seated immediately after all Nations, members of the Council, in the fiftieth seat in the same area'.[39]

B. Rights to be Present, Participate and Exercise Membership Rights

In June 1997, a closed session of the Council was held, during which a situation arose concerning participation of the EC Representative, who was physically barred from entering the Council chamber at some point in the proceedings. The CCLM had previously provided guidance concerning this situation, which is further addressed in paragraph 241 of the Report of the Hundred and Third Session of the Council mentioned above.

Following a review of the applicable legal framework, there appeared to be a need to distinguish between the rights to be present, participate and exercise membership rights. As the FAO Legal Counsel explained:

> At the time, it was agreed that the EC would not be required to leave the Council chamber, or indeed the Council member seating block, on Agenda Items on which

[38] Report of the Hundred and Third Session of the Council, Rome, 14–25 June 1993, paragraph 241 (CL 103/REP).

[39] Report of the Hundred and Third Session of the Council, Rome, 14–25 June 1993, paragraph 242, (CL 103/REP).

the Member States had exclusive competence but would be allowed to continue to be present on those occasions, although not to 'participate' in accordance with Article II.9 of the Constitution.

C. Discussions at the Council

At the Hundred and Sixty-sixth Session of the FAO Council, held on 26 April 2021, a question was raised with respect to the right of the Member Nation holding the EU Presidency to intervene during the Council's discussions on the Chairperson's conclusions and, thus, the report of the Council with respect to agenda item 3, concerning the draft Strategic Framework 2022–31 (C 2021/7).

In particular, the question arose as to whether the Member Nation holding the EU Presidency could participate in deliberations on the conclusions of the Council when that Member Nation is an Observer, and not a Member of the Council.

In response to this situation, a legal opinion issued by the Legal Counsel on 27 April 2021 concluded that

> [w]hile the Independent Chairperson retains authority to accord the right to speak in accordance with the General Rules of the Organization, including to Observers, a review of the practice of the Organization reflects that the Chairperson has consistently permitted an Observer Member Nation holding the European Union Presidency to intervene in the substantive deliberations on agenda items when a Member of the Council has requested this from the Chairperson.

In fact, records show that an Observer representing the Member Nation holding the EU Presidency has been accorded the right to speak on the conclusions of the Council on a number of occasions.[40] In the more recent past, often the Member Nation holding the EU Presidency has also been a Member of the Council.

More generally, the records confirm that practice has been somewhat mixed: other (non-EU) Observers have sometimes been given the floor to comment on the conclusions.[41] However, in other instances, Members of the Council

[40] That is, during the Council Sessions CL149 (June 2014) – Greece; CL151 (March 2015) – Latvia; CL154 (June 2016) – Netherlands; CL155 (December 2016) – Slovakia; and CL156 (April 2017) – Malta.

[41] eg CL155 (December 2016) – Bangladesh and Islamic Republic of Iran; CL165 (December 2019) – New Zealand; CL160 (December2018) – Senegal and Russian Federation; and CL158 (December 2017) – Russian Federation and Jordan.

have questioned whether Observers should have been permitted to intervene on the conclusions of the Council.[42] The Independent Chairperson has, on some occasions, reminded those present that it is the practice that the finalisation of the conclusions of the Council is restricted to the Members of the Council.[43]

D. Participation of the EU at CODEX

In 2003, the EU became a member of CODEX following the adoption by the Commission, at its Twenty-sixth Session, of amendments to the CODEX Rules of Procedure allowing for the admission of REIOs to the membership of CODEX. While the revised CODEX Rules of Procedure[44] mirror the wording of Article II of the FAO Constitution, they are silent with respect to the participation of Member Organizations in the Executive Committee of CODEX (CCEXEC), which falls outside the scope of the definition of subsidiary bodies.[45]

Following a declaration of competence made by the EC in 2003, the EU has not participated as a Member of the Executive Committee and no declaration of competence has been submitted prior to its sessions. However, individuals employed at the European Commission have occasionally attended sessions of the Executive Committee as advisers to the Member for the CODEX region of Europe, which has been the subject of discussions and objections by some Members over the past years.[46]

[42] eg, CL154 (May/June 2016), and CL158 (December 2017).

[43] eg CL155 (December 2016), CL158 (December 2017), CL160 (December 2018), CL165 (December 2019).

[44] Revised CODEX Rules of Procedure, 28th edn (FAO and WHO, 2023), https://www.fao.org/fao-who-codexalimentarius/publications/procedural-manual/en/, Rule II: 'Member Organizations'.

[45] Rule XI.1 of the CODEX Rules of Procedure: 'The Commission may establish the following types of subsidiary bodies: (a) subsidiary bodies which it deems necessary for the accomplishment of its work in the finalization of draft standards; (b) subsidiary bodies in the form of: (i) Codex committees for the preparation of draft standards for submission to the Commission, whether intended for worldwide use, for a given region, or for a group of countries specifically enumerated by the Commission; and (ii) coordinating committees for regions or groups of countries which shall exercise general coordination in the preparation of standards relating to such regions or groups of countries and such other functions as may be entrusted to them'.

[46] Forty-fifth Session of the CODEX Alimentarius Commission, November 2022, Agenda Item 14.1, 'Participation of the European Union as Adviser to the Member for Europe in the Executive Committee of the Codex Alimentarius Commission' (CX/CAC 22/45/21).

Following a review of the legal framework governing the composition of delegations[47] and the related role of advisers,[48] a joint legal opinion from the FAO and WHO Legal Offices concluded that

> the presence of an employee of the European Commission as adviser to the Member of Europe at CCEXEC would not amount *stricto senso* to participation of the EU in a meeting within the meaning of Rule II(2) of the Codex Rules of Procedure.

There is a one unusual particularity concerning voting by Member Organizations under the CODEX legal framework. Decision-making in the Commission is governed by the CODEX Rules of Procedure,[49] complemented by Rule XII of the General Rules of the Organization where matters have not been addressed in those Rules of Procedure.

Rule II.3 of the CODEX Rules of Procedure establishes that

> A Member Organization may exercise on matters within its competence, in any meetings of the Commission or any subsidiary body of the Commission in which it is entitled to participate in accordance with paragraph 2, a number of votes equal to the number of its Member States which are entitled to vote in such meetings and present at the time the vote is taken. Whenever a Member Organization exercises its right to vote, its Member States shall not exercise theirs, and conversely.

Thus, unlike Article II.10 of the FAO Constitution, according to Rule II.3, the number of votes that the EU may cast is limited to the number of its Member States entitled to vote who are actually present in the chamber when the vote is about to take place. This therefore requires that, not only must the number of FAO Members be verified for the purpose of checking whether there is a quorum before proceeding with a vote, but also the number of EU Member States, to confirm how many votes the EU may cast in that particular vote. An example of this voting procedure is reflected in the CODEX Report of the Forty-fifth Session when Members voted on the maximum residue limits for zilpaterol hydrochloride.[50]

[47] FAO Constitution, Article V.1; General Rules of the Organization, Rule III.1; CODEX Rules of Procedure, Rule VI.4.

[48] FAO Constitution, Article V.1; CODEX Rules of Procedure, Rule VI.5 and Rule VIII.1.

[49] See the Procedural Manual of the CODEX Alimentarius Commission.

[50] Report of the Forty-fifth Session of the CODEX Alimentarius Commission, 21–25 November 2022, 12–13 December 2022, and 19 December 2022–6 February 2023, paragraphs 122 and 131. The results of the 'roll-call' vote are contained in Appendix VIII, Part B to the Report (REP22/CAC), https://www.fao.org/fao-who-codexalimentarius/meetings/detail/en/?meeting=CAC&session=45.

VII. The EU as a Strategic Partner and Donor

The EU has made valuable contributions to the work of FAO and has developed a strong partnership with FAO, leading to concrete results in areas including food security, nutrition, fisheries and sustainability at country, regional and international levels. With the adoption of the 2030 Agenda for Sustainable Development, there is a stronger call for greater commitment towards the achievement of the Sustainable Development Goals (SDGs). For FAO, deepening engagement with key partners and fostering alliances are fundamental for eradicating hunger, malnutrition and poverty.

The technical dialogue and cooperation between the EU and FAO began in 2004 with the signature of a Memorandum of Understanding, progressing through close and regular informal meetings, and culminating in the EU–FAO Strategic Dialogue in 2017 with contributions from 10 European Commission Directorates-General.[51] Within this format, the partnership and strategic relations between the EU and FAO are discussed and recalibrated every two to four years, with structured Monitoring Meetings held once a year to chart progress. This enables both partners to strengthen their capacity to bring timely, coherent, and effective responses to global challenges and policy developments within both organisations.

Currently, the EU and FAO are engaged in a Strategic Dialogue with a strengthened focus on the shared goal of eradicating poverty, hunger and malnutrition and jointly advancing the transformation towards more sustainable, efficient and inclusive agrifood systems.

The 2021 edition of the EU–FAO Strategic Dialogue, adopted after the unprecedented global health, social and economic crisis of COVID-19, had the overarching objective of jointly working towards more resilient, sustainable, green and inclusive agrifood systems. The following priorities have been identified for strategic cooperation between FAO and the EU: (i) food systems and 'One Health'; (ii) climate change, biodiversity and natural resources; (iii) food crises, food security and resilient livelihoods; (iv) sustainable agrifood value chain investments and policies; and (v) food system transformation through digitalization.

[51] See FAO, 'FAO + European Union. Investing in a sustainable and food secure future' (Rome 2021), https://web.archive.org/web/20210506151036/http://www.fao.org/3/cb2372en/cb2372en.pdf.

The total approved FAO Budget for 2022–23 is USD 3.25 billion.[52] Of this amount, 31 per cent comes from assessed contributions paid by Member Nations, while 69 per cent is expected to be mobilised through voluntary contributions from Members and other partners.

The voluntary contributions provided by Members and other partners support technical and emergency (including rehabilitation) assistance to governments for clearly defined purposes linked to the results framework, as well as giving direct support to FAO's core work. The voluntary contributions are expected to reach approximately USD 3.5 billion in the period 2022–23.

The EU is one of the largest resource partners in terms of voluntary contributions to FAO, which has enabled FAO to work extensively across the globe and in those regions where assistance is most needed. For the period 2012 to 2017, the EU contributed more than EUR 920 million (USD 1 billion) to projects and programmes implemented by FAO.[53] Most of these contributions were directed towards increasing the resilience of livelihoods to threats and crises; initiatives to make agriculture, forestry and fisheries more productive and sustainable; and helping to eliminate hunger, food insecurity and malnutrition.[54] In the 2020–21 period, the EU contributed approximately USD 334 million, supporting the Organization in providing policy guidance and technical advice, and in implementing programmes and projects in line with the Paris Agreement[55] on climate change and the SDGs.

Within those strategic collaborations, where priorities align, EU resources can serve as a driver and foster innovation to address the most complex international challenges. For example, the EU is an important funder of FAO's Global Programme to support the implementation of the Agreement on Port State Measures to Prevent, Deter and Eliminate Illegal, Unreported and Unregulated Fishing (PSMA)[56] and complementary international instruments, currently contributing around 24 per cent of the USD 27.5 million available to the Programme.

[52] Report of the Conference meeting at its Forty-second Session, Rome, 14–18 June 2021, Resolution 8/2021 concerning Budgetary Appropriations 2022–23.

[53] FAO, 'FAO + European Union. A report on partnership 2016–2017' (Rome, 2018), https://reliefweb.int/report/world/fao-european-union-report-partnership-2016-2017.

[54] See https://www.fao.org/europeanunion/in-figures/en/.

[55] The Paris Agreement was adopted by 196 Parties at COP 21 in Paris on 12 December 2015, and entered into force on 4 November 2016.

[56] The Agreement on Port State Measures to Prevent, Deter and Eliminate Illegal, Unreported and Unregulated Fishing (PSMA) was approved by the FAO Conference at its Thirty-sixth Session under Article XIV.1 of the FAO Constitution, through Resolution 12/2009 dated 22 November 2009. The Agreement was registered with the Secretariat of the United Nations on 26 January 2017 under No I-54133.

Through promoting the adoption and implementation of international instruments, guidelines and tools, this Programme provides capacity development support to states in complying with their flag, port, coastal and market state responsibilities to combat illegal, unreported and unregulated (IUU) fishing. This support includes undertaking legal reviews and drafting to enhance national fisheries legal frameworks, as well as gaps assessment for countries on the implementation of the PSMA and related instruments, and develops response strategies. In practical terms, compliance with PSMA can support compliance with EU rules on IUU. FAO's technical work to support PSMA can help countries in their efforts towards sustainable fisheries.

VIII. Concluding Observations

The unique nature of the EU's status in the Food and Agricultural Organization has enabled active and positive collaboration between the Organization, the EU and the EU Member States. This is underpinned by shared priorities and objectives in a broad range of areas.

In the second half of 2022 and the first quarter of 2023, the EU held discussions aimed at defining the strategic medium-term priorities of the EU and its Member States for FAO. As a result of this exercise, the EU Working Party on International Food and Agricultural Questions reached agreement on the text of the draft Council Conclusions on the EU priorities for an effective FAO to ensure a sustainable future and food security for all.[57] The Council of the European Union considered and approved these conclusions on 20 March 2023.[58]

In the Council Conclusions, the Council of the European Union, inter alia, highlights FAO's vital role 'in leading multilateral efforts to eradicate hunger, all forms of malnutrition and achieving food security and nutrition for all, as well as in tackling poverty'; and acknowledges 'the crucial contribution that FAO is making in the areas of agriculture, including livestock, forestry, fisheries, and aquaculture to build a sustainable future'.[59]

[57] Council of the European Union, 'Conclusions on the EU's priorities for an effective Food and Agriculture Organization of the United Nations (FAO) to ensure a sustainable future and food security for all', Brussels, 10 March 2023, 7222/23.

[58] Council of the European Union, 'Conclusions on the EU's priorities for an effective Food and Agriculture Organization of the United Nations (FAO) to ensure a sustainable future and food security for all', Brussels, 21 March 2023, 7652/23.

[59] ibid, paragraphs 4 and 8.

The Council Conclusions, underline 'the longstanding and fruitful partnership between FAO and the EU and its Member States', recall 'the significant support that the EU and its Member States provide to FAO, including through voluntary contributions', and reaffirm 'the commitment of the EU and its Member States to consolidate and reinforce their cooperation with FAO and to continue working with the FAO membership to ensure effective FAO action for a sustainable future and food security for all'.[60]

Consistent with its status as a specialised agency of the United Nations, FAO will continue to deliver its important technical mandate, in collaboration with its Members and partners – including the European Union – in an independent, impartial and neutral manner. In this context, as the FAO Director-General has stated, the Organization looks forward to continuing its 'close partnership with the European Union, to ensure that FAO remains and further evolves as an efficient, effective and transparent organization that puts people first, thereby accelerating progress towards achieving the Sustainable Development Goals'.[61]

[60] ibid, paragraphs 14 and 16.

[61] FAO, 'FAO + European Union. Investing in a sustainable and food secure future' (Rome, 2021).

4

The Participation of the European Union and its Member States in the International Labour Organization: An Historical Perspective of the EU–ILO Relationship[1]

MAURIZIO BUSSI, MICHAEL BRAUN
AND DAVID MOSLER

I. The Historical Evolution of the EU–ILO Relationship – Three Phases

The partnership between the European Union and the International Labour Organization formally dates to the first cooperation agreement signed by the ILO in June 1958, at the time still with the European Economic Community. Since then, the relationship has strengthened to cover an ever-greater range of employment and social affairs issues, with the EU and its Member States contributing the biggest share to the ILO regular budget, and the EU being the single largest donor to development cooperation efforts of the ILO.

Research conducted by the European Parliament in recognition of the ILO's centenary in 2019 outlined the evolution of the relationship between the EU and the ILO through three distinct phases.[2]

[1] This contribution is based on the intervention made by Maurizio Bussi, ILO Deputy Regional Director for Europe and Central Asia, at the event *EU Diplomacy in Multilateral Fora*, held at the University of Geneva on 31 May 2022.

[2] 'EU and ILO – Shaping the Future of Work' PE 638.407 – June 2019 (europa.eu).

The first phase, ranging from 1958 until 1999 was marked by a more ad hoc approach. At the core was the provision of services to third countries, principally in lower income countries. During that phase, the EU, or its precursor – the European Economic Community (EEC), actively relied on ILO Conventions as a key source of policy advice, particularly on issues such as occupational safety and health, labour mobility and vocational training. While first exchanges were formalised by an exchange of letters as early as 1961, the difference in seniority of the organisations in practice mostly meant that the ILO advised the EEC on issues under its mandate through its established network of international experts.[3] Starting in the 1970s, with the increasing integration of the social dimension, the EEC started to set its own standards and policies but largely ensured their alignment with ILO normative tools. To this end, the Brussels-based bureaucracy expanded its technical expertise substantially.[4] Reflecting the increased weight of EEC legislation in the region, an exchange of letters in 1989 granted the EEC observer status at the ILO.[5] Leading up to the turn of the millennium, the EU became an increasingly active political supporter of international labour standards, culminating in crucial support for the adoption of the ILO Declaration of Fundamental Principles and Rights at Work (1998) and the ILO Convention on the Worst Forms of Child Labour (1999).

Following on the back of these landmark adoptions of ILO tools, the EU entered a new phase of much closer collaboration with the ILO in the years 2000–07. It became sort of a policy incubator for the ILO, using normative approaches within EU Member States for potential use across different regions and in support of the ILO Decent Work Agenda. During this time, the two institutions also established a system of high-level meetings among senior representatives of the organisations as a way to formalise the collaboration on a more structured basis and signed their first cooperation agreement in 2000. This agreement formally established the attendance of one institution at relevant meetings of the other, high-level meetings between the organisations and formal contact points to facilitate exchange. The results of deeper collaboration showed, for example, in the very active role the European Commission took in the World Commission on the Social Dimension of Globalization, which was established by the ILO in 2002.

[3] A Johnson, 'Social policy at the global level: EU Member States at the ILO', European Union Studies Association (EUSA), Biennial Conference 2003 (8th), 27–29 March 2003, Nashville, TN.

[4] A Johnson, 'EU–ILO relations: Between regional and global governance', in J Orbie and L Tortell (eds), *The European Union and the social dimension of globalisation. How the EU influences the world* (London, Routledge, 2009) ch 5.

[5] Exchange of Letters between the European Communities and the International Labour Organization (90/C 24/06) (21–22 December 1989), https://eur-lex.europa.eu/legal-content/EN/TXT/PDF/?uri=OJ:JOC_1990_024_R_0008_01.

Since 2008, the ILO and the EU have actively collaborated through reciprocal consultations on political issues but have also created much closer links for continuous cooperation on various technical areas of expertise, and the ILO started providing more assistance to EU Member States as well. This effectively made the two institutions equal partners.[6] The Global Financial Crisis and even more so the Eurozone Crisis of the early 2010s were a catalyst for bringing about change in the relationship. During the Eurozone crisis in particular, labour market challenges such as high unemployment, the exclusion of youth and women, and shortcomings in social protection systems came to the forefront, and the ILO increasingly engaged with EU Member States to offer technical assistance and advance the knowledge frontier.[7] The ILO now even advised EU Member States in Southern Europe to minimise negative social and employment effects of the crisis itself, as well as assisting with the responses articulated under the respective Memoranda of Understanding that said countries signed with the so-called Troika.[8] A crucial element of ILO engagement in these countries was support for social dialogue mechanisms, which were severely curtailed by some of the Memoranda. As the crisis nearly led to some countries breaking with the EU, the presence of the ILO served as an assurance, particularly to social partners, whose trust in the EU (and often in their national governments) had severely diminished.

While the sovereign debt crisis in these countries has passed, this deeper engagement continues to serve as the basis for EU–ILO collaboration, which more recently has focused on the challenges posed by the Fourth Industrial Revolution,[9] as the EU engaged strongly in deliberations on the future of work leading up to the ILO's centenary celebrations in 2019 and resulted in the adoption of the ILO Centenary Declaration for the Future of Work during the 108th session of the International Labour Conference.

II. ILO–EU Relations and Joint Work

Much of the collaboration by the two organisations outside of the EU concerns the joint promotion of common values. This was most recently reaffirmed in a

[6] EC Landau and Y Beigbeder, *From ILO Standards to EU Law: The Case of Equality between Men and Women at Work* (Leiden, Martinus Nijhoff Publishers, 2008).

[7] See, for example, the series of labour market analyses entitled 'Studies on Growth with Equity' produced by the ILO and International Institute for Labour Studies.

[8] The Troika consisted of the European Commission, the European Central Bank and the International Monetary Fund.

[9] Namely, the current era of connectivity, advanced analytics, automation, and advanced-manufacturing technology that has transformed global business in the 21st century.

2021 exchange of letters which highlights the priorities for cooperation, including a human-centred approach to the future of work and a just transition to a future of work that contributes to sustainable development.[10] Broadly, cooperation can be divided along four main axes: (i) the EU's support to advance the standard-setting agenda of the ILO, (ii) the integration of international labour standards and technical advice from the ILO into EU policy frameworks, (iii) the EU's financial and technical support for ILO capacity-building efforts in third countries, and (iv) research collaboration in key areas of interest to advance the knowledge frontier.

Often these four axes intersect, as can be seen in the concrete example of the reciprocity between the EU's and the ILO's agendas concerning the digital transformation of work in the form of the ILO's Centenary Declaration for the Future of Work, and some of the various regulatory initiatives taken by EU institutions in the context of the Pillar of Social Rights. The EU was a crucial partner working towards the adoption of the Centenary Declaration; it included central elements of the ILO agenda, such as the Decent Work Agenda, in the Pillar of Social Rights and invited the ILO to contribute and comment during its development. Prior to the adoption of the Pillar of Social Rights at the 2017 Gothenburg Summit, and also prior to the reaffirmation and the adoption the accompanying action plan at the 2021 Porto Summit, the ILO was closely engaged with the EU to embed the long-elaborated legacy of ILO standards in this important policy framework.

The EU is also a strong partner in terms of ratification of ILO Conventions, of which there are 190 at the moment, although some have, over time, been abrogated or withdrawn. EU Member States had ratified 2226 Conventions and Protocols by May 2022, accounting for more than a quarter of the 8410 ratifications at global level by the 187 Member States of the ILO.[11] Due to the subsidiarity principle, the EU is not necessarily in a position to determine whether or not its Member States should ratify certain ILO legal instruments, since much of the competence for social and employment policy is within the remit of national governments. Nevertheless, there are informal attempts at coordinating and aligning EU legislation in Member States to reflect provisions and obligations under ILO standards, and the European Commission usually gives a recommendation to Member States on the ratification of new standards.

[10] Exchange of Letters between the ILO and the European Commission (4 February 2021), https://www.ilo.org/publications/exchange-letters-between-european-commission-and-international-labour.
[11] Data derived from NORMLEX – Information System on International Labour Standards, as at 31 May 2022.

Ratification and application of ILO standards is also an important subject for countries aspiring to join the EU. With chapter 19 of the EU Acquis Communautaire covering social policy and employment, helping potential members with compliance has been a long-standing effort of the ILO, particularly through its sub-regional office for Central and Eastern Europe, in Budapest. Most recently, this has involved the roll-out of Youth Guarantees in some candidate countries, which has been financially supported by the EU through the funding of ILO technical assistance.

Moreover, the EU is a passionate promoter of social dialogue and tripartism, both internally and externally. Social dialogue is a fundamental component of the European social model. It enables the social partners (representatives of employers and workers) to contribute actively, including through agreements, to the design of European social and employment policy. Social dialogue is, of course, one of the key foundations of the ILO, which is the only multilateral organisation in which governments, employers and workers come together.

EU and national social partners are consulted in many legislative processes, particularly through public consultations on social policies that are standard practice for European Commission proposals. The so-called European social partners – umbrella organisations of national social partners in EU Member States – are engaged in European social dialogue, provided for under Articles 154 and 155 of the Treaty on the Functioning of the European Union (TFEU), and engage in an annual EU Social Summit with the European Council and the European Commission. The method of social dialogue for decision-making in the EU has been strengthened through interaction with the ILO.

In external policy, effective tripartite social dialogue has become an increased focal point for the EU. For example, the establishment of tripartite social dialogue practices at national level in Georgia was the result of the EU's political pressure, the ILO's technical assistance and the International Trade Unions Confederation's support. A nod to social dialogue also exists in the adoption of the Fundamental Principles and Rights at Work as part of all newly-signed EU free trade agreements since 2008. Since February 2022, as established by the European Commission's Communication on Decent Work Worldwide, the has EU aimed to embed the principles of Decent Work more generally in its foreign policy and development assistance.[12] In the spirit of this Communication, European Commission

[12] European Commission, Communication on decent work worldwide for a just transition and a sustainable recovery, Brussels, COM(2022) 66 final. https://eur-lex.europa.eu/legal-content/EN/TXT/HTML/?uri=CELEX:52022DC0066.

President, Ursula von der Leyen, announced a proposal for a ban on entry into the EU for products made with forced labour in late 2022, capitalising on the market-making role of the EU as a large single consumer market. A similar spirit drove the adoption of the proposal for a Directive on corporate sustainability due diligence. This mix of policy tools makes clear reference to ILO standards and shows the growing importance that the EU attaches to ensuring Decent Work at home and abroad.

Of course, the EU also plays a crucial role as a primary donor to ILO development cooperation. The EU is the largest donor to the ILO's development cooperation programme, having provided USD 445 million in voluntary contributions between 2014 and 2021, and currently supports 65 ongoing projects. While the thematic breadth of supported projects is vast, there are some key areas of EU interest where financial support is particularly strong. These include Decent Work in global supply chains; universal social protection; employment impact assessments of trade and investment; economic empowerment of women at work; implementation of ILO fundamental Conventions; skills development, employability and youth employment; and labour migration.

The EU is an important political actor in the ILO's institutional organs. While only its individual Member States have voting rights in ILO organs, namely the Governing Body and the International Labour Conference, the EU often plays a role of coordination and facilitation among EU Member States. EU Member States have issued common statements in the ILO since 1973. They demonstrate declaratory cohesion (issuing common statements) for the same instruments as they demonstrate voting cohesion. Since the EU itself is used to building consensus and reaching compromise between its Member States, it is often able to rally ILO Member States around issues and causes.

The EU also played a key role in cooperating closely with emerging economies, developing countries and social partners, for example during the adoption of the June 2008 ILO Declaration on Social Justice for a Fair Globalization, during the adoption of the Maritime Labour Convention in February 2006, and during the adoption of the Global Jobs Pact in 2009.

This role of the EU as a catalyst for multilateral diplomacy has evolved over time, with the EU increasingly becoming an important multilateral player. In 2021, the European Commission issued a Joint Communication to the European Parliament and the Council on 'strengthening the EU's contribution to rules-based multilateralism'. This states that the EU wants to promote global peace and security and defend the fundamental rights, universal values and international law that form the bedrock of cooperative relations between countries and peoples. It will remain at the forefront of the response to the existential climate and biodiversity crises that threaten humanity, and which can only be addressed through strong multilateralism.

This renewed engagement is driven by multiple factors. Of course, geopolitical competition plays a role, but EU policy-makers were also some of the earliest advocates for global solutions to global problems and realised the EU's responsibility, as one of the largest consumer markets in the world, for the working and living conditions in countries along the global supply chains feeding European consumption. Clearly, the global COVID-19 pandemic also served as a catalyst towards multilateral solutions, as it demonstrated the weakness of non-coordinated national responses, as well as the intricate links between a global health event, supply chains, workers' occupational safety and health, labour migration and many other factors. As a result, the EU promotes cooperative solutions to 'build back better' – strengthening the global recovery, tackling inequalities, promoting sustainable development and public health, enabling the digital and green transitions including clean energy transition, and defending human rights and the rule of law.

These efforts go hand-in-hand with a more interests-based approach. This is to be achieved by reinvigorating the multilateral order with the UN at its core, to make it fit for purpose. It calls for the EU to engage actively in support of the UN Secretary-General's 'Our Common Agenda' – the vision of the UN Secretary General for the UN until 2046, when the centenary of the UN will be celebrated.

III. Institutional Interactions between the EU and the ILO Today

The main cooperation partner among EU institutions for the ILO is the European Commission, not least through its role as a donor to technical cooperation projects. In recent years, the ILO also increasingly cooperates with EU institutions other than the Commission.[13]

The ILO is regularly referenced in European Parliament debates and texts, ILO representatives are sometimes asked to participate in hearings or meetings at the Parliament and the ILO Office in Brussels follows several parliamentary committees as part of its engagement and advocacy work. For example, in preparatory work for the EU Minimum Wage Directive adopted in 2022, ILO technical

[13] The relevant agreements with the various institutions can be found at: https://www.ilo.org/european-union-eu.

experts were asked by parliamentary political groups to give input on compliance with the relevant international labour standards.

The European Council explicitly supports Decent Work and the social dimension of globalisation and has made reference in multiple summit statements to the centrality of Decent Work.

The Council of the European Union, working under different configurations of line ministers dependent on subject-matter, firmly supports the promotion of Decent Work for all, reinforcing the social dimension of globalisation and the promotion of fundamental social rights at work. The ILO regularly participates as an observer in meetings of the Employment, Social Policy, Health and Consumer Affairs Council (EPSCO) – the group of line ministers responsible for social and employment policy.

The European Economic and Social Committee (EESC) is a consultative body of the EU and consists of representatives of employers, workers and other civil society organisations. In 2005, the EESC and the ILO signed a joint declaration on the strengthening of relations and cooperation.

Social partner organisations at European level include in particular the European Trade Union Confederation (ETUC) and Confederation of European Business (BusinessEurope). Other employers' organisations are the Association of Crafts and SMEs in Europe (SMEunited) and the European Centre of Employers and Enterprises Providing Public Services (CEEP). These organisations, through their participation in global umbrella organisations, also impact decisions at the ILO's Governing Body and the International Labour Conference, often being a driving voice behind negotiations.

IV. Case Study: Free Trade Agreements and International Labour Standards

In the area of core labour standards, the ILO has reinvented itself as the primary and unique universal institution advocating for fundamental principles and rights at work (FPRW), which were first adopted with the ILO Declaration on Fundamental Principles and Rights at Work in 1998 and were amended to include a fifth element in 2022. These principles and rights consist of:

- freedom of association and the effective recognition of the right to collective bargaining;

- the elimination of all forms of forced or compulsory labour;
- the effective abolition of child labour;
- the elimination of discrimination in respect of employment and occupation; and
- a safe and healthy working environment.

These are considered universal values with the same weight as international human rights. The EU has supported the ILO in efforts to promote them universally, not just through giving technical assistance in countries around the world, but also by including them in EU trade agreements since 2008. The 2021 exchange of letters between the European Commission and the ILO states:

> As regards trade policy, the promotion of Decent Work, ratification and implementation of core labour standards and occupational safety and health have become part of EU trade agreements with, and preferences granted by the EU to its partner countries and an important area of cooperation between both organizations.[14]

This has been an important tool to promote FPRW globally, but also helps the EU to level the playing-field by making it harder for third countries to increase their competitive advantage in production through workers' rights abuses.

The applicable Regulation for the EU's Generalised Scheme of Preferences (applying to both GSP and GSP+ countries) allows for the EU to withdraw tariff preferences in a number of circumstances, eg, in the event of serious and systematic violations of the core 15 UN and ILO Conventions by beneficiaries. One instance when this sanction mechanism was applied was in 2020, when the European Commission decided to partially withdraw Cambodia's preferential access to the EU market based on concerns about human rights and labour rights.[15] The Commission proposal for a new Regulation for the EU's GSP system – commencing in 2024 and approved in late 2022 – reinforces this mechanism. Increased enforcement mechanisms, including a phased approach on withdrawal, are certainly a trend.

A cursory account of labour provisions within EU free trade agreements tells a positive story. Trade agreements have been negotiated with relatively extensive

[14] Exchange of Letters between the ILO and the European Commission (2021), n 10 above.

[15] Commission Delegated Regulation (EU) 2020/550 of 12 February 2020 amending Annexes II and IV to Regulation (EU) No 978/2012 as regards the temporary withdrawal of the arrangements referred to in Article 1(2) of Regulation (EU) No 978/2012 in respect of certain products originating in the Kingdom of Cambodia, OJ L 127, 22.4.2020, pp 1–12.

substantive standards and procedural commitments. But in terms of addressing substantive labour standards issues, work remains to be done, in particular in relation to enforceability. EU trade agreements generally do not include provisions on trade remedies and/or sanctions for non-compliance with or failure to implement trade and sustainable development provisions (with the exception of the EU–UK trade agreement). The debate about non-compliance with labour standards is a very sensitive one and the ILO has always stayed neutral on the question in exchanges with EU institutions, which is in line with the guidance given by the Governing Body on trade and labour provisions in 2016. Of course, the ILO's supervisory mechanism continues to adjudicate on infractions of international labour standards, but from the perspective of the individual countries concerned, rather than through their trade relations.

While remaining neutral, the ILO has always advocated that progress cannot be achieved without dialogue, engagement, ownership and cooperation. The ILO always indicates that it is available to provide any support deemed useful for promoting dialogue, engagement and follow-up corrective action. ILO Director-General Gilbert Houngbo has indicated that the ILO stands ready to accompany partners in their trade negotiations should they so desire.

In this context, the European Commission is currently working on a Trade Policy Review, possibly considering reinforcing the sustainability chapters in its free trade agreements, which have already been bolstered in the last iteration in 2021. The ILO recently submitted a technical note during the public consultation process. The note outlines the general observations on the labour provisions of EU trade agreements and identifies possible areas where engagement in promoting Decent Work in the future can be further expanded. The note does not make recommendations on a phased approach in case of non-compliance but focuses on the need for continuous dialogue and building the capacities of constituents to ratify and implement international labour standards, which requires technical cooperation.

To further enhance the knowledge base on labour provisions in trade agreements, the ILO recently launched the ILO Labour Provisions in Trade Agreements Hub, the development of which was funded by the European Commission and the Government of Flanders. It is a comprehensive, structured compilation of the text of labour provisions in more than 100 bi- and pluri-lateral regional trade agreements. It also includes information on the provisions in trade agreements remedies for non-compliance (such as compensation, suspension of benefits, monetary contributions, compliance/action plans and other measures, and border access restrictions). The hub is a compilation and offers possibilities to ILO constituents and researchers for analysis, including when advising numerous governments during the negotiations of new trade agreements.

To give more teeth to labour provisions, one of the elements that could be considered is to strengthen the time-bound nature of the commitments undertaken in free trade agreements or GSPs. It would be helpful if agreed action plans or road maps had some timeline. This may not be possible with respect to detailed engagement, but for example, it would be much easier to ensure follow up on ratification if the terms of the agreement already indicated that ratification would take place by a certain date. In the case of the free trade agreement with South Korea, the 'continued and sustained efforts' requirement is very vague, and it is unclear at what point such efforts can no longer be raised.

We might see an expansion of the dialogue between WTO and ILO in the future. Building on the Centenary Declaration, the Global Call to Action for a Human-Centred Recovery from the COVID 19 crisis states that:

> the ILO will strengthen cooperation with relevant multilateral and regional organizations and processes to achieve a strong and coherent global response in support of national recovery strategies, including in order to: … coordinate Decent Work objectives and capacity-building assistance more closely with international trade and investment policies to widen the benefits of international trade and investment and promote Decent Work, environmental sustainability and sustainable enterprises in supply chains, taking into account the strong, complex and crucial links between social, trade, financial, economic and environmental policies.[16]

The EU plays a crucial role as an institution strongly represented in both institutions. Multilateral coherence is one of the hallmarks of EU foreign policy, and it clearly starts with also ensuring coherence between different policy portfolios within the EU and its Member States.

V. Concluding Remarks

The relationship between the EU and the ILO has evolved over the decades. While initially, the ILO was the more established institution, advising the newly-founded EEC, European integration steadily changed the relationship, turning the two into

[16] ILO, 'Global call to action for a human-centred recovery from the COVID-19 crisis that is inclusive, sustainable and resilient' (Geneva, October 2021), https://webapps.ilo.org/digitalguides/en-GB/story/globalcall#intro.

equal partners. In the early days, much of the cooperation was driven by individuals in both organisations, seeing peaks and troughs of activity. Since the 1970s, cooperation has become much more formalised, with regular participation in joint meetings and continuous technical exchanges. Today, the ILO continues to provide its specialised expertise to support new legislative efforts in the EU, but the EU also plays a much stronger role in the normative function of the ILO. Moreover, the ILO benefits from generous EU-funding for many of its development cooperation projects, which are seen as a win for both organisations – enabling the ILO to help countries in complying with international labour standards, while encouraging EU foreign policy that supports human rights as both a moral imperative and an economic effort to protect its own industries from unfair competition. The collaboration has established Decent Work – a core ILO concept – as a central element of EU internal and external policy and turned the EU and its Member States into active drivers of new standard-setting at the ILO.

5

'*Wer rastet, der rostet*': Towards a Greater Role of the European Union in the Food and Agriculture Organization, World Health Organization and International Labour Organization

FRANK HOFFMEISTER

I. Introduction

Wer rastet, der rostet, is a German proverb indicating that a person who takes a break may also take on some rust. In a certain way, this may be also relevant for the EU, when looking at its role in several international organisations.

In the 1990s, the former European Community became a full member of the Food and Agriculture Organization (1991) and the WTO (1994). Following a policy push from the Commission in 2003,[1] the Community also advanced its status in other multilateral fora from a mere observer to an active participant in some UN bodies in the first decade of the new century.[2] After the entry into force of the Lisbon treaty in 2009, the European Union secured active participant rights in the UN General Assembly in 2011.[3] However, further progress, as envisaged by

[1] Communication from the Communication to the Council and the European Parliament – *The European Union and the United Nations: The choice of multilateralism*, COM(2003)526 final.

[2] F Hoffmeister, 'Outsider or Frontrunner? Recent Developments under International and European Law on the Status of the European Union in International Organisations and Treaty Bodies' (2007) 44(1) *CML Rev* 41, at 55.

[3] UNGA Res 65/276, Participation of the European Union in the Work of the United Nations, 3 May 2011.

the Barroso–Ashton strategy paper of 2012[4] relating to the EU's status in international organisations, remained minimal. The paper was probably not focused enough, and in any case, the Commission leadership at the time did not mobilise sufficient political energy in the Council and beyond.[5]

In 2021, the High Representative and the Commission together reviewed the status quo again. In their joint communication on strengthening multilateralism they focused on the EU's potential to contribute to the standard-setting and policy-making of international organisations. The EU's status as such was less relevant, as long as it would not hamper its ability to make its voice known in a given multilateral forum. In this respect, they observed:

> In many instances and for various reasons, the EU does not have full membership rights of international organisations. This has not prevented formal arrangements in the form of participation as observer or even pragmatic ad hoc arrangements with Member States and/or the organisations concerned and their members, providing the EU and its representatives with the capacity to effectively intervene and act beyond its formal status. The EU is recognised as a reliable constructive partner and a deal maker. A recent example was the May 2020 World Health Assembly where, despite not being a member, the EU was the driving force behind the Resolution on COVID-19 response. Nevertheless, in other important cases such arrangements either do not exist, are unnecessarily cumbersome, outdated, or prevent the Commission and/or the High Representative from speaking effectively on behalf of the EU in matters falling under EU competence. One case in point is the Food and Agricultural Organisation, where the EU's capacity to be effective is undermined by outdated arrangements with Member States that are not in line with the Treaties. The EU should therefore urgently assess on a case-by-case basis whether its status in a given international organisation, fund or programme requires adaptation or upgrade, in accordance with the EU Treaties, and do what is necessary to achieve the adaptation or upgrade.[6]

Moreover, the Communication mentions the International Labour Organization (ILO) as an important international standard-setter, which the EU wishes to support as follows:

> Economic recovery also goes hand in hand with social justice, social protection and decent work. International labour standards set out in the International Labour

[4] Communication to the Commission from the President in Agreement with Vice-President Ashton – *Strategy for the progressive improvement of the EU status in international organisations and other for a in line with the objectives of the Treaty of Lisbon*, C(2012) 9420 final, 20 December 2012.

[5] For discussion of the strategy and its limits, see J Wouters, A Chané, J Odermatt and T Ramopoulos, 'Improving the European Union's Status in the United Nations and the UN System: An Objective Without a Strategy?', in C Kaddous (ed), *The European Union in International Organisations and Global Governance: Recent Developments* (Oxford, Hart Publishing, 2015), 45, 73–74.

[6] European Commission and High Representative, Joint Communication to the Council and the European Parliament, *Strengthening the EU's contribution to rules-based multilateralism*, JOIN(2021) 3 final, 17 February 2021, 11–12.

> Organisation provide a tried-and-trusted foundation for inclusive recovery but more can be done to promote worldwide ratification of important ILO conventions. A well-functioning world economy also requires multilateral rules, adapted to the realities of the 21st century that allow for open, fair and rules-based trade, and ensure a level playing field for all economic actors. Keeping markets open, avoiding an escalation of trade conflicts and modernising trade rules are critical for a fair and sustainable recovery.[7]

This remark hints at another interesting new development, namely how the EU's trade policy may become an important 'enabler' for stronger ratification and implementation of ILO Conventions.

The present contribution will hence look at the EU's institutional role in the FAO and the WHO more closely and identify some more recent developments about the EU's indirect strengthening of the ILO via its 'sustainability chapters' in Free Trade Agreements with third states. For that purpose, the EU's status and its role in policy-making in each of the three organisations will be discussed before concluding.

II. The EU at the Food and Agriculture Organization

Although Rome is no less attractive than Geneva as a location, the EU's full membership in the FAO usually attracts far less academic attention than the EU's presence in the WTO.[8] Nevertheless, the status of the EU within the FAO raises an increasing number of interesting institutional questions (for the lawyer) and policy questions (for the diplomat).

The starting point is the EU's power to conduct a comprehensive agricultural and fisheries policy, today enshrined in Articles 38–44 TFEU. Characterised as a shared competence (Article 4(2)(d) TFEU), the adoption of EU legislation in the field has the consequence that Member States cannot exercise their powers any more in the same area (Article 2(2) TFEU and Protocol 25 to the Lisbon Treaty). Moreover, a part of the fisheries policy, namely the conservation of marine biological resources, even falls under EU exclusive competence. That point was first made by the European Court of Justice in 1976,[9] and is now codified in Article 3(1)(d) TFEU.

[7] ibid, 5.

[8] See F Hoffmeister, 'The EU in the World Trade Organization – A model for the EUs Status in International Organisations?', in C Kaddous (ed), *The European Union in International Organisations and Global Governance: Recent Developments* (Oxfort, Hart Publishing, 2015), 121–40.

[9] ECJ, Joined Cases 3/76, 4/76 and 6/76 *Kramer and others*, ECLI:EU:C:1976:114, paras 30–33.

Taken as a whole, the EU's powers in the agricultural and fisheries domain have grown considerably over time with every new piece of EU legislation. Moreover, the EU's environmental policy under Article 191 TFEU (another shared competence according to Article 4(2)(e) TFEU) produces additional legislation with relevance for matters arising in FAO.[10]

A. Legal Status of the European Union

(i) Historical Evolution

As far back as 1961, the then Commission President of the European Atomic Energy Community and the Director-General of FAO agreed on a structured relationship between them. Their joint interest was focused 'on the use of radioisotopes and radiations in research, processing and production in the field of food and agriculture including forestry and fisheries' and on the 'the study of the movement, behaviour and concentration of radionuclides in soils, water, plants, animals and food and their relevance to human diet'.[11] Roughly 30 years later, such narrowly focused cooperation was replaced by a completely different approach. Taking into account the European Economic Community's ever growing agricultural and fisheries policy as described above, FAO members agreed after more than two years of negotiations since the submission of the EEC's accession request in 1989 to allow full membership via a constitutional amendment in November 1991.

However, that did not mean automatic EU membership in all FAO-related bodies or Conventions negotiated under the auspices of the organisation under Article XIV of the FAO Constitution. That *lacuna* became politically evident when the WTO Appellate Body emphasised in 1998 the significance of the *Codex Alimentarius* rules under Article 2.4 of the Technical Barriers to Trade (TBT) Agreement and Article 3 of the Sanitary and Phytosanitary Measures (SPS) Agreement in the widely noted Hormones dispute between the US/Canada and the EU.[12] Hearing the wake-up call,[13] the EU then intensified its efforts to be

[10] For the delimitation between the EU's fisheries and environmental policy see ECJ, Case C-683/16 *Deutscher Naturschutzring v Germany*, ECLI:EU:C:2018:36.

[11] Exchange of letters between the President of the Euratom Commission and the Director-General of FAO of 9 December 1961, OJ L 43 7.6.1962, p 1356.

[12] WT/DS48/AB/R. Report of the Appellate Body of 16 January 1998.

[13] B Van der Meulen and B Wermaart, 'Food and Agriculture Organization (FAO) and Codex Alimentarius Commission', in RA Wessel and J Odermatt (eds), *Research Handbook on the European Union and International Organizations* (Cheltenham, Edward Elgar Publishing, 2019), 92–93.

admitted as a full member also to the *Codex*, a joint subsidiary body of the FAO and the WHO established in 1961. After lengthy discussions and a vote in the *Codex Alimentarius* Commission (73 for, 12 against, 13 abstentions), the EC also became a member there in 2003. A year later, in 2004, the European Commission and the FAO Director-General signed a 'Strategic Partnership', and in 2005 the EC also became a party to the International Plant Protection Convention.[14] In 2014, a new Memorandum of Understanding replaced the 2004 document, and in 2017 and 2021, two successful EU–FAO Strategic Dialogues took place.

The EU and its Member States are also an important donor for FAO. For the 2024–25 period, less than 30 per cent of FAO's expenses are covered by the regular contributions of its members. The other 70 per cent of the roughly 2 billion USD annual FAO budget for these years stems from voluntary contributions. The EU usually pays 150–200 million USD in a given year, granting support for FAO programmes and projects.

(ii) FAO Rules

Article II(4)–(10) FAO Constitution regulate the EU's participation in the FAO as a 'Member Organization'. It thus enjoys the right to participate and the right to vote, the details of which can be summarised as follows.

(a) The Right to Participate in Governing Bodies

FAO's internal structure is rather complex. According to a decision of the FAO Conference, the Governing Bodies are defined as follows:[15]

> The Governing Bodies comprise the Conference, the Council, the Programme Committee, the Finance Committee, the Committee on Constitutional and Legal Matters, the Technical Committees referred to in Article V, paragraph 6(b) of the Constitution and the Regional Conferences (i.e. for Africa, Asia and the Pacific, Europe, Latin America and the Caribbean, and the Near East).

In this list, the FAO Conference and the four technical committees are open to all FAO members. Accordingly, the EU is also admitted to these plenary organs as

[14] Council Decision 2004/597/EC, OJ L 267, 14.8.2004, p 37.

[15] FAO Conference (Rome, 17–14 November 2009), C 2009/REP, para 142. FAO *Basic Texts*, Vol II, section B, https://openknowledge.fao.org/server/api/core/bitstreams/437ae43e-7426-4e48-9d99-623dde284258/content#page=137.

a Member Organization. For the Regional Conferences, it is logical that the EU takes part in the Regional Conference for Europe.

The situation is more complicated for the Council as the executive organ of the Conference and its Committees. Article II(9) of the FAO Constitution reads:

> Except as otherwise provided in this Article, a Member Organization shall have the right to participate in matters within its competence in any meeting of the Organization, including any meeting of the Council or other body, other than bodies of restricted membership referred to below, in which any of its Member States are entitled to participate. A Member Organization shall not be eligible for election or designation to any such body, nor shall it be eligible for election or designation to any body established jointly with other organizations. A Member Organization shall not have the right to participate in bodies of restricted membership specified in the rules adopted by the Conference.

Two rules can be deduced from this text. First, the EU cannot be present in restricted bodies, which have been identified in relevant rules adopted by the Conference. Currently, this is the case for the General Committee of the Conference, the Credentials Committee and the Nominations Committee. Moreover, the EU is not admitted to any Committee established by the FAO Council under Article V(6)(a) of the FAO Constitution (ie the Programme Committee, the Committee on Constitutional and Legal Matters, and the Finance Committee).

Second, in the FAO Council itself (consisting of 49 members elected by the Conference), the EU can be present in its sessions. However, it shall not 'participate' on matters where exclusive competence remains with the Member States.[16] In such scenario, the EU representative may remain 'present' in the chamber, but cannot take the floor. When the agenda item concerns shared competence, the EU position is usually expressed by the Member State holding the rotating Presidency of the EU Council (irrespective of whether or not that EU Member State is an elected member of the FAO Council or just an observer).

(b) The Right to Participate in Statutory Bodies

Next to the Governing Bodies, the FAO also operates Statutory Bodies. While the former owe their existence to a decision of the FAO Conference or the FAO Council, the Statutory Bodies are established under Articles VI and XIV of the FAO Constitution. Today, there are over 100 special bodies, which deal with

[16] FAO Council, Report of the 103rd Session, Rome, 14–25 June 1993, paras 237–42 (CL 103/REP).

precise individual subject-matters falling under FAO's mandate (eg forestry or desertification, etc). Whenever it has competence in the subject-matter, the EU can fully participate in them.

A difficult issue arose, though, with the *Codex Alimentarius* Commission, established together with the WHO in 1961. Could the EU become a member of this joint body, given that it was a Member Organization of the FAO, but only an observer in the WHO? As observed above, the question was resolved in favour of the EU only in 2003, when it became a full member of the *Codex*.

The final category are bodies established under a Convention adopted by FAO according to Article XIV of the Constitution. Here, the EU may participate if it has become a party to the Convention itself, which in turn depends on the scope of the EU's competence for ratifying that specific Convention. This is the case, for example, for certain regional fisheries management organisations, for which the EU holds exclusive competence.

(c) Voting Rights

Turning to voting, Article 2(10) of the FAO Constitution provides:

> Except as otherwise provided in this Constitution or in rules set down by the Conference, and Article III paragraph 4 notwithstanding, a Member Organization may exercise on matters within its competence, in any meeting of the Organization in which it is entitled to participate, a number of votes equal to the number of its Member States which are entitled to vote in such meeting. Whenever a Member Organization exercises its right to vote, its Member States shall not exercise theirs, and conversely.

This system requires a clear understanding of the EU's competence in a given matter. In practice, the FAO not only asked for a general declaration on competence under Article II(5) of the FAO Constitution upon the EU's accession in 1991, but also for a precise indication of the division of competence as between the EU and its Member States before any vote. EU Voting on elective[17] and budget issues[18] is excluded.

The weight of an EU vote is 'equal to the number of its Member States', ie currently 27. A rather strange exception exists, though, within the *CODEX Alimentarius* due to the resistance of one *Codex* member (the United States)

[17] FAO General Rules of the Organisation, Rule XLV(2).
[18] Article XVIII(6) of the FAO Constitution.

during negotiations.[19] Although the Codex Rules largely follow Article II(10) of the FAO Constitution,[20] Article II(8) of the Codex Rules mandates that the EU vote is dependent on the Member States being actually present in the meeting. This is a legally unsatisfactory arrangement because Member States long ago vested competence in the Union by concluding the Treaty: they do not grant ad hoc 'empowerments' through their presence.[21] In practice, this nuance does not diminish the EU's influence in *Codex* votes, however, as the EU will organise full Member State presence for all important meetings.

(iii) EU Rules

While the FAO rules define the participation and voting rights of the EU within the organisation, the EU needs to decide internally how a position is brought about and how it is represented in FAO meetings. In order to settle these questions, the Council and the Commission agreed back in December 1991 an 'arrangement', which was slightly updated in 1992 and 1995.[22] The main thrust works as follows: all EU positions are prepared in the FAO Working Party or other relevant working parties of the Council, such as the one on fisheries and forestry. The FAO Working Party meets roughly once a month in Brussels and also decides about the competence issue. The positions are either proposed by the Commission, or a Member State or the Council Presidency. They are adopted by unanimity in practice.

When the agenda item refers to a matter of exclusive EU competence, the Commission will speak. When the agenda item falls within Member States' competence, the Member States themselves will speak. If, however, an agenda item falls within shared competence, there will be a 'common position' and the thrust of the matter becomes decisive to determine the EU representative. If the thrust of the matter lies within exclusive EU competence, the Commission will present the common position in the FAO, and if this is not the case, the Member State holding the Presidency of the Council delivers the common position of the

[19] J Maersk Pedersen, 'FAO-EU Cooperation: An ever stronger partnership', in J Wouters, F Hoffmeister and T Ruys (eds), *The United Nations and the European Union* (The Hague, TMC Asser Press, 2006) 63, 84.

[20] For details see Van der Meulen and Wermaart, 'Food and Agriculture Organization (FAO) and Codex Alimentarius Commission', above n 13, 93–95.

[21] See critique by J Sack, 'The European Community's membership of international organizations', (1995) 32 *CML Rev* 1237, 1241 with respect to similar arrangements in certain old commodity agreements.

[22] Council Document 10478/91 of 18 December 1991; Council Document 9050/92 of 7 October 1992; Council Document 8460/95 of 26 June 1995.

EU and its Member States. In case of disagreement between the Commission and the Council, the matter is brought to the level of the EU's Committee of Permanent Representatives (COREPER) and the Council, whose decisions are subject to judicial review before the Court of Justice as the arrangement codifies the primary law duty of sincere cooperation.[23]

With the entry into force of the Lisbon Treaty, the external representation of the EU was fundamentally reformed. According to Article 17(1) sixth sentence TEU, (only) the Commission represents the EU in areas of TFEU policies, whereas Common Foreign Affairs and Security Policy positions are for the High Representative to present (Article 27(2) TEU). EU Delegations in third countries and international organisations can speak on behalf of the EU across the entire spectrum (Article 221(1) TFEU). Accordingly, the Presidency of the Council has no role anymore in representing the EU – at most, it can be asked by Member States to represent them in an area of (pure) Member State competence. This new system was then put into a 'General Arrangement', adopted by the Council in October 2011,[24] inter alia, in order to overcome a blockage caused by the United Kingdom about the proper formula to be used post-Lisbon. That document underlined that only 'EU actors' can make statements on behalf of the EU – and the corresponding footnote clarifies that such actors are 'the President of the European Council, the High Representative, the European Commission and the EU Delegations'.

For dealings with the FAO, though, the General Arrangement contained the following exception in the last indent of point 3:

> EU actors will conduct local coordination and deliver statements on behalf of the EU unless prevented by the rules of procedure of the forum concerned (default setting). Where practical arrangements such as those at the World Trade Organisation, at the Food and Agricultural Organisation and in burden sharing exist for coordination and/or representation, such arrangements will be implemented for the preparation and delivery of the statement on behalf of the EU from behind the EU nameplate.

The reason for keeping the pre-Lisbon FAO arrangements alive is not set out in the General Arrangement. Probably, there was no appetite in the Council to touch the established practice in the relevant working party and in the EU Delegation

[23] See ECJ, Case C-25/94 *Commission v Council*, ECLI:EU:C:1996:114 with detailed commentary from R Terreiro Teixeira and J Wouters, in Butler and Wessel (eds), *EU External Relations Law – the Cases in Context* (Oxford, Hart Publishing, 2022) ch 31, 329–38.

[24] Council Document 15901/11 of 24 October 2011, at https://data.consilium.europa.eu/doc/document/ST-15901-2011-INIT/en/pdf, Annex.

in Rome. Moreover, as mentioned above, it was usually the rotating Presidency of the Council, who would take the floor for the EU in the FAO Council.

While this might be understandable for a certain transitional period after the entry into force of the Lisbon Treaty in December 2009 to ensure an efficient performance in Rome, it became less convincing over time. Hence, the Commission proposed an update in 2013,[25] but failed. It seems that a majority of Member States continued to believe that the rotating Presidency of the Council has an important role to play in the FAO field. This comes at a double cost: first, the system of preparing EU positions in Brussels stays as rigid as it was for the previous 30 years, leaving very little room for adapting EU positions in Rome; second, the EU Delegation cannot play the role it has assumed under Article 221 TFEU, in a marked contrast to the role it plays in Geneva and New York, where the EU is not even a member of the respective UN bodies and specialised agencies! Instead, maintaining a central role for the rotating Presidency risks jeopardising the institutional memory and the stability for external EU contacts.

Against that background, the 2021 Communication has brought the issue of the outdated 1991 arrangements back to the table. The Commission and the High Representative then presented a revised draft arrangement to the Czech Presidency in the second half of 2022. Under the draft, the main coordination work would shift to Rome and only EU actors would safeguard external representation of EU positions. At the same time, the system of burden-sharing and the special expertise of Member States in particular areas would remain in place and no EU actor could qualify to take on elective functions.

This draft is currently being discussed at the Working Party for International Food and Agricultural Issues, with input from the Working Party on UN Matters (CONUN) and other relevant working parties. Clearly, purely legal reasoning will not bring about a decisive switch in the Council's approach. However, more and more Member States may realise that it is time to strengthen the EU as a whole within the FAO given the latter's increasingly political role, to which we now turn. In other words, if FAO matters are not seen merely as technical agri-fish matters, left to experts from the Ministries of Agriculture, but also as important foreign policy for Foreign Ministries, there may be a better chance that the Council as a whole realises the benefit of a stronger role for the EU Delegation in Rome.

[25] COM(2013) 333 final.

B. Political Role of the European Union

What are these recent political changes at the FAO where the EU role is more acutely felt? Importantly, in March 2023, the Council of the EU defined the EU's priorities 'for an effective FAO to ensure a sustainable future and food security for all'.[26] The document touches upon FAO's role in addressing the acute food crisis (points 6–7), FAO's role in building a sustainable future (point 8), FAO's rule in the humanitarian-development–peace nexus for addressing food insecurity (point 9), and FAO's role in the wider UN System (point 10). It also sets out EU expectations to improve the organisation's effectiveness and governance (points 11–13) and reiterates the 'longstanding and fruitful partnership between FAO and the EU and its Member States', looking forward to an 'enhanced dialogue and collaboration' (points 14–16).

While it is impossible to deepen all these points in this contribution, some remarks on three of them may shed some light on the EU's current political approach towards FAO.

(i) Food Security in the Aftermath of the Russian Invasion of Ukraine

FAO is an important partner for the EU in the area of food security, mainly because of its capacity to exert impact on the ground with its numerous field offices worldwide. Thus, the EU has increased its cooperation with FAO since 1991. The projects have mostly related to African countries and concern sustainable agricultural policies and statistical cooperation.[27] This area was hence characterised by an increased engagement on both sides up to 2022.

In the aftermath of the Russian invasion of Ukraine in February 2022, food security became also a political priority. Since Russia and Ukraine make up one-third of global cereal exports and many countries depend on imports from either country, a quick and convincing policy response was needed, also to counter quickly rising prices. Surprisingly, though, FAO Director-General Qu Dongyu on 15 March 2022 published an opinion piece,[28] which ignored the fact that the General Assembly had condemned the Russian aggression in a resolution of 2 March 2022 and just referred to a 'challenge' that there might be no Ukrainian

[26] Council Document 7652/23 of 21 March 2023, Annex.

[27] Maersk Pedersen, 'FAO–EU Cooperation: An ever stronger partnership', n 19 above, 63, 69–76, and 78–82 with concrete examples.

[28] FAO, News Article: 'New Scenarios on Global Food Security based on Russia-Ukraine Conflict'.

harvest in June that year. The Director-General also found that financial sanctions against Russia 'have caused an important depreciation which, if continued, could undermine productivity and growth and ultimately further elevate agricultural production costs'. Clearly, the implication that Western sanctions were a root-cause of the global food crisis, while ignoring the real cause, caused concern about FAO's policy analysis.

Based on an initiative from France and the United States, an extraordinary meeting of the FAO Council was then scheduled for 8 April 2022. During that session, the FAO Council expressly reaffirmed the UNGA resolution and asked the FAO Secretary-General, inter alia, to develop immediate, medium-term, and longer-term plans to address the global food security impacts of the conflict in Ukraine, including emergency response, recovery assistance, and resilience-building in food systems.[29] A month later, in a presentation for the G7 of 13 May 2022,[30] the Director-General then brought forward a more balanced analysis. He identified three types of risk deriving from the ongoing war:

- Food and Agriculture (input, harvest, trade, logistics, prices);
- Macro, ie financial risk associated with macroeconomic, social or political factors (energy, debt, nuclear contamination);
- Humanitarian (food, migration and refugees).

Accordingly, he proposed a FAO Rapid Response Plan for Ukraine, suggesting how to maintain food production and supply chain in the country, combined with a global Food Import Financing Facility for vulnerable countries (which led to the International Monetary Fund establishing a 'Food Shock Window').

At the end of the year, the issue came back to the FAO Council. On the initiative of the US, and supported by the EU and like-minded states, the FAO Council adopted another decision in relation to the effects on global food security of the Russian aggression in Ukraine.[31] During the deliberations, the FAO Council first voted down the request of the Russian Federation for an amendment, then supported the text from the like-minded states,[32] and finally rejected a Russian

[29] FAO Council Decision of 8 April 2022 on Impact of the Ukraine-Russia conflict on global food security and related matters under the mandate of the Food and Agriculture Organization of the United Nations (FAO), Report of the 169th session, 8 April 2022, CL/169/Rep, p 8, para 11(e)(iii).

[30] Presentation for the Director-General Meeting of the G7 Ministers of Agriculture (fao.org).

[31] See FAO Report on the 173rd session (Rome, 5–9 December 2022), CL 173/Rep, para 12; https://www.fao.org/3/nl148en/nl148en.pdf.

[32] See FAO Council Doc CL173/3, Add. 1, https://www.fao.org/3/nk954en/nk954en.pdf.

(counter-) decision[33] with clear majorities.[34] Both the convening of the extraordinary Council session in April 2022 and the fact that the December 2022 Decision was adopted by vote (instead of the usual consensus in FAO) showed the increased political attention of the EU and like-minded states for the organisation.

At the same time, FAO cannot take over the political role of the main UN organisation relating to peace and security or of the World Food Programme to tackle humanitarian emergencies. Against such broad understanding of a division of labour between the UN, WFP and FAO (concentrating on mid- and long-term food policies), a 'big issue', namely how to guarantee further Ukrainian exports, quickly left the organisation. The 'Black Sea Grain Initiative' was negotiated under the auspices of Turkey and the UN Secretary-General. It was first agreed on 23 July 2022, renewed in November 2022 and March 2023. The arrangement allowed the export of over 23 million tons of grain and other foodstuffs via three Ukrainian ports,[35] which helped in addressing the global food crisis and lowering grain prices.

Against that background, the Council made the following political assessment in its March 2023 conclusions. The Council was

> (6) deeply concerned about the unprecedented food crisis the world is facing, affecting in particular the poorest countries and most vulnerable populations, and thereby deepening inequalities; reiterate[d] its resolute condemnation of Russia's unjustified and unprovoked war of aggression against Ukraine, whose negative and long lasting impacts on food security are severely exacerbating the crisis;
>
> (7) reaffirm[ed] the EU's strong commitment to respond to global food insecurity and reiterate[d] its full support for the UN-led efforts to avert further escalation; recognise[d] FAO's efforts to monitor and address the consequences of Russia's war of aggression for food security in Ukraine and globally *and urge[d] FAO to intensify those efforts, fully implementing the relevant decisions by FAO governing bodies* (emphasis added).

(ii) Regulatory Standards for Food Safety

In the area of setting global standards for food safety, the *Codex Alimentarius* occupies centre stage. As mentioned above, the EU became a full member in 2003. It is hence able to influence the standard-setting activity. At the same time, the EU is also a faithful follower. Article 5(3) of the EU's General Food Law

[33] See FAO Council Doc CL173/3, Add. 2, https://www.fao.org/3/nk955en/nk955en.pdf.
[34] See FAO Council, Procès Verbal of the 173th session, https://www.fao.org/3/cc3924t/cc3924t.pdf, pp 54–63.
[35] See https://www.consilium.europa.eu/en/infographics/ukrainian-grain-exports-explained/.

(Regulation 178/2002) calls upon Member States to take into account international standards where they exist or their completion is imminent; as a result, EU food safety standards and *Codex* standards are mostly aligned on all major topics.[36] It hence seems that this area is governed by a satisfactory 'business as usual'.

(iii) Fisheries Management

Another important area is fisheries management, mentioned in point 8 of the March 2023 Council Conclusions. Back in 2001, FAO had adopted an International Plan of Action on 'illegal, unreported and unregulated fishing' (IUU fishing).[37] It contained a set of voluntary measures for flag states and coastal states to prevent, deter and eliminate such practices. Even at the time of its adoption, the EU argued for a tighter system[38] and subsequently took the lead by imposing a comprehensive EU legal framework in 2008.[39] In its fight against overfishing in the high seas, the EU established a system of inspection of third-country vessels in EU ports and a restrictive trading system.[40] Importantly, the EU also determines whether a third country is cooperating in the fight against IUU practices. After a warning (the 'yellow card') the EU may put a third country on a black-list (the 'red card'). Such listing triggers an import prohibition for any fisheries product of that country, associated with a prohibition of EU vessels operating in its waters (Article 38 of the 2008 IUU Regulation). In practice, according to a European Parliamentary report of 2021, the Commission has identified more than 50 non-cooperating states in the last decade, out of which six received the red card. Three of them were delisted after 13, 20 and 35 months, respectively.[41] In 2021, eight countries continued to have a yellow card, and three suffered from total import prohibitions due to a red card. These figures imply that the other 40+ countries reacted positively to the EU concerns.

[36] Van der Meulen and Wermaart, 'Food and Agriculture Organization (FAO) and Codex Alimentarius Commission', n 13 above, 96–98.

[37] https://www.fao.org/documents/card/en/c/71be21c9-8406-5f66-ac68-1e74604464e7.

[38] Report of the Twenty-fourth Session of the Committee on Fisheries (Rome, 26 February–2 March 2001)' FAO Fisheries Report No 655 para 98, https://www.fao.org/3/y2161e/y2161e.pdf.

[39] Council Regulation (EC) 1005/2008, of 29 September 2008 establishing a Community system to prevent, deter and eliminate illegal, unreported and unregulated fishing OJ L 286, 29.10.2008, p 1.

[40] For details and a discussion of WTO compatibility of the IUU trading regime see F Hoffmeister, 'The European Approach on Supply Chain Responsibility' (2022) *Zeitschrift für europarechtliche Studien* 221, 227–29.

[41] European Parliament Research Service, 'Illegal, Unreported and Unregulated Fishing', Infographic of March 2021, p 2.

In view of such strong EU action, FAO members also increased their attention paid to IUU fishing. Clearly inspired by the EU's 2008 IUU Regulation, the FAO Conference adopted in 2009 an Agreement on Port State Measures barring access to ports for vessels engaged in IUU fishing. After its entry into force in June 2016, FAO agreed in 2017 to launch a 'Global Record' of fishing vessels, refrigerated transport vessels and supply vessels. This important tool keeps an eye on all relevant activities and helps to eliminate or deter IUU fishing globally.

One modality for 'illegal' fishing comes into play when a vessel violates the quotas enacted by one of the five regional fisheries management organisations (RFMOs). As the EU is a party to the instruments establishing them, it is also an important player in running them. One recent example may illustrate the important political interaction between the RFMOs, FAO and the EU. It relates to the Indian Ocean Tuna Commission (IOTC), which is open to all FAO members, and of which the EU is a member as well.

According to Article IV(1) of the Establishment Agreement of the IOTC, a member can join either as a 'coastal State' (ie having sovereign territories around the Indian Ocean), or as a 'State or Associate Member whose vessels engage in fishing in the Area for stocks covered by this Agreement' (ie not being coastal State, but having an active fishing interest in the area). Since the deposition of its instrument of accession in 1995, the UK has been a member of the IOTC as a coastal State on account of Chagos as its 'British Indian Ocean Territory' (BIOT). Mauritius has systematically raised the issue of sovereignty over the Chagos archipelago, contesting the membership of the UK as a coastal State. The UK also considered itself covered in its fishing nation capacity as part of the instrument of acceptance deposited by the EU until the end of the transition period, ie 31 December 2020.[42]

In its Advisory Opinion delivered on 25 February 2019,[43] the International Court of Justice concluded, inter alia, that 'the United Kingdom is under an obligation to bring to an end its administration of the Chagos Archipelago as rapidly as possible'. The UN General Assembly in its resolution of 22 May 2019[44] called upon all the specialised agencies (including FAO) 'to recognize that [Chagos] forms an integral part of the territory of Mauritius' and to refuse to recognise any reference to 'BIOT'.

[42] See Article 129 of the Withdrawal Agreement of the UK from the European Union, OJ CI 384, 12.11.2019, p 1.

[43] ICJ, *Legal Consequences of the Separation of the Chagos Archipelago from Mauritius in 1965*, Advisory Opinion, ICJ Reports 2019, p 95.

[44] UNGA Res 73/295 of 24 May 2019, Advisory opinion of the International Court of Justice on the legal consequences of the separation of the Chagos Archipelago from Mauritius in 1965.

In December 2020, just before leaving the European Union, the UK resubmitted its accession instrument to the IOTC, without, however, specifying whether the UK submitted it as a 'coastal State' or as a 'fishing nation'. The FAO Legal Counsel in her opinion of 6 May 2022 concluded that it is legally unacceptable to interpret that instrument as submitted by the UK in its capacity of a coastal State.[45] As a consequence, the UK can only qualify under IOTC rules as a 'fishing nation'. In their Twenty-sixth annual meeting in 2022, the IOTC Parties discussed the issue without a resolution.

According to the jurisprudence of the European Court of Justice,[46] the EU is bound to respect customary international law, including the right to self-determination, as interpreted by the International Court of Justice. Additionally, at the conclusion of the EU–UK Trade and Cooperation Agreement, a Union declaration was issued on 1 June 2021, according to which: 'The European Union declares that the reference to the British Indian Ocean Territory in paragraph 4 of Article 774 of the Agreement (1) is to be interpreted and implemented in full respect of applicable international law'.[47] Accordingly, taking into account the ICJ Opinion and the opinion of the FAO Legal Counsel, the accession instrument of the UK has to be interpreted in the sense that the UK can be member of the IOTC exclusively as a State whose vessels engage in fishing in the area under Article IV(1)(a)(ii) of the IOTC Establishment Agreement.

C. Conclusion

On the one hand, the EU at the FAO is a credible Member Organization. On the other hand, it has not yet achieved its full potential. Crippled by the resistance of some Member States to allowing the EU delegation in Rome to negotiate for and to represent the EU in the organisation, the Union's political influence punches below its weight. In particular, in view of the recent attention paid to FAO's role in securing global food security after Russia's invasion of Ukraine, the EU should become more

[45] Note of 6 May 2022 from the FAO Legal Counsel in relation to Agenda item 4 of the agenda for the 26th Session of the Indian Ocean Tuna Commission, proposed by Mauritius: 'Clarification of the Status of the United Kingdom's Membership in the IOTC', at https://iotc.org/sites/default/files/documents/2022/05/IOTC-2022-S26-12ENote_from_the_FAO_Legal_Counsel_regarding_agenda_item_4E.pdf.

[46] ECJ, Case C-266/16 *Western Sahara Campaign UK v Commissioners for HM Revenue & Customs*, ECLI:EU:C:2018:118; Case C-104/16 P *European Council v Front Polisario*, ECLI:EU:C:2016:973; and Case C-363/18 *Organisation Juive Européenne and Vignoble Psagot v Ministre de l'Économie et des Finances*, ECLI:EU:C:2019:954.

[47] https://eur-lex.europa.eu/legal-content/EN/TXT/?uri=CELEX%3A22021X0601%2801%29.

politically astute in Rome. Together with the overdue need to adapt to the Lisbon arrangements, it is therefore high time to upgrade the role of the EU Delegation in Rome to help increase the European influence in the organisation overall.

III. The EU at the World Health Organization

According to Article 9 TFEU, the Union shall take into account the protection of human health in defining and implementing its policies. Accordingly, whenever the EU legislates in the internal market, it shall observe a high level of health protection (Article 114(3) TFEU). This explains, for example, a number of EU Directives on tobacco products and labelling. According to Article 168 TFEU, the Union may also support the policies of the Member States, but not occupy the field itself by legislative means, except in narrowly described circumstances. The EU can hence legislate under Article 168(4) TFEU on the quality and safety of donated human tissues and cells; on organs and substances of human origin; on veterinary and phytosanitary matters which have as their direct objective the protection of public health; and on the quality and safety of medical products and devices.[48] For those safety concerns in public health matters, the powers are hence shared between the EU and its Member States (Article 4(2)(k) TFEU), with the consequence that EU legislation may prevent Member States from enacting laws in the same field (Article 2(2) second sentence TFEU).

One important means of EU financial support is the Health Programme for 2021–27, as laid down in Regulation (EU) 2021/522 with a budget of over €5 billion,[49] adopted under Article 168(5) TFEU. Finally, Article 35 of the Charter of Fundamental Rights enshrines the right of access to preventive healthcare and the right to benefit from medical treatment 'under the conditions established by national law and practices'.

A. Legal Status of the European Union

Membership in the WHO is open to states only (Article 3 WHO Constitution). At the same time, the Constitution offers three possibilities for engagement with

[48] For a detailed analysis of the EU's powers in health policy see E Ruiz Cairó, *The promotion of public health in EU external relations* (Geneva/Zurich, Schulthess, 2021), 51–74.
[49] Regulation (EU) 2021/522, OJ L 107, 26.3.2021, p 1.

other intergovernmental organisations.[50] The first option involves the Director-General, who may 'establish direct relations with international organizations whose activities come within the competence of the Organization' (Article 33 WHO Constitution). Under the second option, the World Health Assembly (Article 18(h) of the WHO Constitution) may invite participants to assist in its work without the right to vote. The third option is to conclude a formal agreement between the WHO and UN specialised agencies (Article 69 WHO Constitution) or other international organisations (Article 70 WHO Constitution).

(i) General Legal Status of the EU in the WHO

For their general interaction, the EU and the WHO have chosen the first option so far. Following an exchange of letters of 1972 with the WHO Regional Committee of Europe, the European Communities and the WHO Director-General defined arrangements for their cooperation by an exchange of letters of 1982.[51] It foresaw the reciprocal participation in meetings and the exchange of information. Point 3 read:

> 3. The World Health Organization shall invite a *representative of the European Communities* to participate as observer in sessions of the World Health Assembly, the Executive Board and the Regional Committee for Europe and, where appropriate, in the work of their committees, with respect to items on their agenda in which the European Communities and the World Health Organization have a common interest (emphasis added).

However, the subsequent letter from 2000[52] no longer spoke about the 'European Community' (ie the organisation as such), but contained a 'Memorandum concerning the framework and arrangements for cooperation between the World Health Organisation and the *Commission* of the European Communities' (emphasis added).

In view of that changed language, the question arises whether the Commission or the EU as a whole acts as an observer at the WHO. Recently, the European Court of Justice examined a similar constellation at the International Maritime Organization (IMO), where the Commission had signed an exchange of letters with

[50] GL Burci, 'The European Union and the World Health Organization: Interactions and Collaboration from a Governancen and Policy Perspective', in C Kaddous (ed), *The European Union in International Organisations and Governance: Recent Developments* (Oxford, Hart Publishing, 2015), 155, 157–58.

[51] OJ L 300, 28.10.1982, p 20.

[52] OJ C 1, 4.1.2001, p 7.

the Secretary-General of the IMO in 1974. The Court rejected the Commission's argument that it had acted on behalf of the EU – with the consequence that the EU as such is not considered to be an observer in the IMO.[53]

Nevertheless, the 2000 exchange of letters based on Article 33 of the WHO Constitution did not change the WHO practice of inviting the European Community[54] (and subsequently the EU) to its meetings. By default, the EU is 'equated to'[55] an international organisation with which the Organization has established effective relations under Article 70 of the Constitution. Consequently, the EU is treated as an observer in the Health Assembly, Executive Board and Regional Committees.[56] Pursuant to the respective rules of these three bodies,[57] it may thus participate without eligibility to vote in open meetings of the bodies themselves and their committees. However, it can only take the floor after Members, non-Member States and observers with special status such as Palestine. Moreover, with the exception of the Executive Board, where the EU representative sits next to the President at the main table, it is seated at the back of the room with the other observers.[58]

The EU Delegation in Geneva coordinates the substance of EU statements and positions in the WHO with the health experts of Member States and organises the outreach process. According to the Council General Arrangement of 2011, an EU actor should then represent the EU in the WHO – which should be either the Commission, the High Representative or EU Delegation. At the same time, positions on public health rarely fall entirely or predominantly within EU competence, but usually involve mainly the competence of Member States. The latter can then be represented by the rotating Presidency, who would then speak on behalf of both the EU and its Member States. Moreover, in cases of doubt as to where the thrust of an EU Member State position lies, it plays a role that an EU actor can only speak as an observer from the back-row. Therefore, in practice, often the Presidency

[53] ECJ, Case C-161/20 *Commission v Council*, EU:C:2022:260, paras 54–57.

[54] B Eggers and F Hoffmeister, 'UN-EU cooperation on public health: the evolving participation of the European Community at the WHO', in J Wouters, F Hoffmeister and T Ruys (eds), *The United Nations and the European Union* (The Hague, Asser, 2006) 155, at 158.

[55] Burci, 'The European Union and the World Health Organization', above n 50, at 162.

[56] L Chamorro, 'Coordination between the European Union and the Member States: An EU Perspective', in C Kaddous (ed). *The European Union in International Organisations and Governance: Recent Developments* (Oxford, Hart Publishing, 2015), 142, 146.

[57] Rule 47 of the Rules of Procedure of the Health Assembly, Article 4 of the Rules of Procedure of the Executive Board, Article 2 of the Rules of Procedure of the Regional Committee for Europe.

[58] T Emmerling, 'World Health Organisation (WHO) and other global health bodies – The EU voice in a fragmented global health landscape', in R Wessel and J Odermatt, *Research Handbook on the European Union and International Organisations* (Cheltenham, Edward Elgar, 2019), 120, 123.

rather than the EU Delegation represents the EU in the World Health Assembly or the Executive Board.[59] Due to the constraints imposed by its observer status, the EU hence exercises still an 'almost purely pre-Lisbon' arrangement for external representation in the WHO Governing Bodies.[60]

(a) Negotiation of WHO International Legal Instruments

An important upgrade from this status exists when the WHO negotiates international treaties which touch upon EU competence. In such intergovernmental meetings, which take place between the WHO Governing Bodies' sessions, the second option became operational: the EU received special invitations from the World Health Assembly with respect to the Framework Convention on Tobacco Control (1999–2003)[61] and the International Health Regulations (2003–05)[62] to participate as a 'regional economic integration organisation'. It was hence able to make proposals and to reply to the positions of others, with the sole restriction of not being allowed to vote.[63] The 'REIO' clause was borrowed from multilateral environmental conventions in the UN system, to which the EU is a party and 'avoided the sensitivities surrounding a special status for the EU as such'.[64] Over time, the initial resistance of third states to allowing the EU Delegation to be seated next to the Member States holding the EU Council Presidency also disappeared. By now, the EU Delegation sits next to the rotating Presidency in such negotiations.

The same arrangement of full participation also applies for the ongoing WHO negotiations on an international treaty on pandemic preparedness and prevention initiated by the World Health Assembly on 1 December 2021. In its decision establishing an intergovernmental body, it expressly invited also 'regional economic integration organisations, as appropriate'[65] to participate. Under the established 'REIO' arrangement, the EU Delegation and the rotating Presidency can hence work hand-in-hand during the treaty negotiations.

[59] Burci, 'The European Union and the World Health Organization', above n 50, at 161.
[60] Chamorro, 'Coordination between the European Union and the Member States: An EU Perspective', above n 56, at 149.
[61] Decision of the World Health Assembly WHA 52.38 of 24 May 1999, para 1(3).
[62] Decision of the World Health Assembly WHA 56.28 of 28 May 2003, para 2.
[63] For details of these negotiations see Eggers and Hoffmeister, 'UN-EU cooperation on public health: the evolving participation of the European Community at the WHO', above n 54, at 162–68.
[64] Burci, 'The European Union and the World Health Organization', above n 50, at 163.
[65] Decision of the World Health Assembly (WHA) of 1 December 2021, SSA2(5), point 1(1), fn 3.

(b) EU Status in WHO International Legal Instruments

When it comes to the conclusion of such WHO instruments, a distinction must be made. Under the WHO's Framework Convention on Tobacco Control (FCTC) as a self-standing international treaty, the EU was also allowed to become a party as a REIO (Articles 34, 35 FCTC), which it did in 2004.[66] The EU is hence present in the FCTC bodies and can exercise alternative voting depending on the competence at hand (Article 32(2) FCTC).

The situation is more complicated with the International Health Regulations (IHR). As they are enacted by the World Health Assembly as binding secondary law under Articles 21 and 22 of the WHO Constitution,[67] the EU cannot be a party to them. So when the decision was taken in the World Health Assembly, the EU Council of Ministers authorised the Member States to vote in favour also in the interest of the EU. The EU is hence bound to fully apply the IHR.[68] However, Article 57(3) IHR also states: 'Without prejudice to their obligations under these Regulations, States Parties that are member of a regional economic integration organization shall apply in their mutual relations the common rules in force in that regional economic organization'. It follows that the EU's IHR obligations apply vis-à-vis third states, whereas EU law applies in the relations between the Member States in the fields covered by EU competence.

(c) EU Participation in Drafting Groups

Finally, WHO practice has become more flexible also with respect to the EU's participation in Drafting Groups convened during the sessions of the World Health Assembly or the Executive Board. According to the second sentence of Rule 47 of the WHA[69] and Rule 4 of the Executive Board, respectively, such participation is dependent on a specific case-by-case invitation. In January 2006, the resistance of some non-WHO members to allow EU participation in such drafting groups was overcome. After a request from Portugal in the opening session of the Hundred and seventeenth Executive Board meeting, the Board accepted EU participation in

[66] Council Decision 2004/513/EC, OJ L 213, 15.6.2004, p 8.

[67] For details on the legal nature of WHO regulations see Burci and Vignes, *World Health Organization* (The Hague, Kluwer 2004), 131–32.

[68] Eggers and Hoffmeister, 'UN-EU cooperation on public health: the evolving participation of the European Community at the WHO', above n 54, at 168, fn 29; Emmerling, 'World Health Organisation (WHO) and other global health bodies', above n 58, at 122.

[69] Rules of Procedure of the World health Authority, https://apps.who.int/gb/bd/PDF/bd47/EN/rules-of-procedure-en.pdf.

drafting groups during that session dealing with technical items falling in whole or in part under EU competence. However, following an intervention of the US and Australia, it was also clarified that the EU Member States would then not take the floor on the same matter, and this would not set a precedent for other international fora.[70] Ever since, this arrangement set a pattern that has been followed not only in the Executive Board, but also for the technical committees of the WHA.[71] It means that at the beginning of each session of a Governing Body, an EU Member State requests that the EU Delegation may participate in such WHO drafting groups, which is usually granted by consensus.[72]

(d) Full Participant Status

In December 2022, the Commission unveiled the new 'EU Global Health Strategy'. In Guiding Principle 14, it enumerated its vision 'to support a stronger, effective and accountable WHO'. The first point is to 'Seek formal EU observer status with full participation rights as a first step towards full WHO membership'.[73] This seems to be a hint at using the third option, which the WHO has used so far with the *Organisation internationale des epizooties* and the African Union.[74] The reasoning behind seems to be two-fold. First, the current arrangement under Article 33 of the WHO Constitution and the 'ad hoc requests' to participate in Drafting Committees to be made for each session of the WHA and the Executive Board may seem a bit fragile and cumbersome. This rather weak *de jure* status[75] could be replaced with a formal arrangement providing more stability. Second, any positive experience with such a future arrangement could muster further support of non-EU WHO states for full EU accession at a later stage.

In this respect, Article 70 of the WHO Constitution provides:

> The Organization shall establish effective relations and co-operate closely with such other inter-governmental organizations as may be desirable. Any formal agreement entered into with such organizations shall be subject to approval by a two-thirds vote of the Health Assembly.

[70] See WHO Executive Board Report EB117/2006/Rec/2, pp 3–4.

[71] Burci, 'The European Union and the World Health Organization', above n 50, at 166.

[72] Emmerling, 'World Health Organisation (WHO) and other global health bodies', above n 58, at 123.

[73] COM(2022) 675 of 30.11.2022, p 15.

[74] For the agreement between the WHO and the African Union see the Resolution of the World Health Assembly WHA65 of 26 May 2012, in WHA65/2012/Rec/1, p 47.

[75] Emmerling, 'World Health Organisation (WHO) and other global health bodies', n 58, at 124.

While the procedure for granting observer status on the WHO side is clear (two-thirds approval by the WHA), the same cannot be said for the European side. Were the EU to ask for full membership, this would entail an amendment to the WHO Constitution, as Article 3 thereof restricts membership to states only. In order to modify this provision by an amendment, the Council would have to issue negotiation directives to the Commission under Article 218(3) TFEU and later on sign/conclude the amending agreement under Article 218(5) and (6) TFEU.

However, asking for an EU full participant status could be based on Article 220(1) TFEU as a sort of 'special relation' with the WHO as a specialised agency of the UN.[76] According to Article 220(2) TFEU that article is 'implemented' by the High Representative and the Commission. Given that health policy falls under the TFEU and that Article 168(3) TFEU calls upon the Union to foster cooperation with competent international organisations in the sphere of public health, the Commission would hence be in the lead. It would have to act in close consultation with the High Representative as he carries the responsibility for the EU Delegation in Geneva, which is also maintaining the relations with the WHO. Once the Commission and the High Representative were to agree on the details for such a status request for the EU, they would have to consult with the Council before officially submitting the request to the WHO on behalf of the European Union.

A positive outcome of such a formal agreement between the WHO and the EU could be to model EU speaking rights along the lines of those granted by the UN General Assembly resolution of May 2011. It could remove the restriction of speaking 'last' and allow the EU to participate more visibly in the work of the WHO by EU actors. For that, the Rules of Procedure of the WHO Governing Bodies would have to be adapted as well.

B. Political Role of the European Union

The EU's role in WHO policy making has increased since the adoption of the EU Global Health Strategy in 2010.[77] The Council announced the EU vision on the link between health policy and sustainable and equitable development in some detail, which was largely aligned with WHO priorities. A concrete outcome was the adoption of the FCTC Protocol on Illicit Trade adopted in 2013 under

[76] Hoffmeister, 'Outsider or Frontrunner', above n 2, at 59–60 with respect to Article 302 TEC.
[77] Council Conclusions on the EU role in Global Health of 10 May 2010, Council Doc 9505/10.

EU influence and leadership[78] (the EU even chaired the negotiations for that legal text).[79] Moreover, the EU invested part of its research and investment program into innovation for global health (healthcare technologies, social innovation), health security as well as health equity.[80]

The EU Global Health Strategy 2022 is equally comprehensive. For the purpose of this contribution, it is appropriate to focus on the EU priorities in dealing with the WHO itself.

(i) Financing of the WHO

Among the EU priorities is to put WHO resources on a more stable basis. Currently, the Commission provides a big chunk of voluntary contributions to WHO programs, and the EU and its Member States together make up to 20 per cent of the overall budget. However, despite being the biggest single contributor, the political return for the EU is not equivalent, and the WHO lacks the necessary legal certainty for long-term planning. So, it would seem to create a win-win situation if a different mode of WHO financing were to be found in line with the decisions of the seventy-fifth World Health Assembly by 2030–31.[81]

(ii) WHO Governance

Another EU focus is directed towards governance, efficiency, accountability of the organisation and the power to enforce its decisions. From a European perspective, the WHO's response towards the Ebola crisis revealed a couple of shortcomings, in particular with respect to the organisation's ability to lead concerted international efforts in taming the spread of the disease.[82] In the COVID-19 crisis, the WHO's limits in developing a global vaccination policy also became apparent. In the end, the EU rolled out its own dissemination plan. While it led to an incredibly high proportion of 86 per cent of vaccination inside the EU with highly

[78] Burci, 'The European Union and the World Health Organization', above n 50, at 169–70 with further details.

[79] Emmerling, 'World Health Organisation (WHO) and other global health bodies', above n 58, at 124.

[80] See in detail I Kickbusch, SA Matlin, S Battams and A Jahn, 'The European Union's Role and Potential in Research and Innovation for Global Health', in C Kaddous (ed), *The European Union in International Organisations and Governance: Recent Developments* (Oxford, Hart Publishing, 2015), 176, at 191–216.

[81] WHA Decision 75(8).

[82] Emmerling, 'World Health Organisation (WHO) and other global health bodies', above n 58, at 133.

effective drugs,[83] the access of other countries to them remained considerably lower.

(iii) EU Support for the WHO as Norm-setter

Finally, the EU aims at strengthening the WHO's focus on its normative role in areas of global relevance. In the past, the two organisations were already convergent in their respective responses to cross-border health threats. As mentioned above, the EU participated in the negotiations of the International Health Regulations of 2005 and influenced their content. It also plays an important role in its implementation.[84] According to an administrative memorandum of 2011 between the WHO and the Commission, there is regular exchange on communicable disease surveillance and control,[85] and the Commission brought together alert mechanisms at international and EU levels by revising Decision No 1882/2003 establishing the European alert system EWRS.[86] However, given the limits detected during the combat of COVID-19 a revision of the IHR is necessary from an EU point of view.

In the same vein, a specific new agreement on combating pandemics should be concluded by 2024. In March 2022, the Council of the EU authorised the Commission to negotiate the instrument on behalf of the EU in the area of EU competence,[87] setting out the EU's objectives in these negotiations. The Union will strive for an ambitious outcome with an increased surveillance and coordinated international response mechanism. Crucially for Europe, the instrument should strengthen global supply chains for medical equipment and provide incentives for research and innovation.

In addition, new topics, such as health data governance and workforce would also be appropriate priorities for world-wide regulation from an EU point of view. In turn, the convergence achieved between EU norms on healthy diets and the WHO Global Strategy on Diet, Physical Activity and Health, adopted by the World Health Assembly in 2004,[88] does not call for a legally binding multilateral convention.[89]

[83] https://www.consilium.europa.eu/en/policies/coronavirus/covid-19-research-and-vaccines/.
[84] See COM(2006) 552 final.
[85] Ruiz Cairó, *The promotion of public health in EU external relations*, above n 48, at 181.
[86] Decision 1082/2013, OJ L 293/1, 5.11.2013 (see recital 16).
[87] https://www.consilium.europa.eu/en/press/press-releases/2022/03/03/council-gives-green-light-to-start-negotiations-on-international-pandemic-treaty/.
[88] WHA Resolution VHA57.17 of 22 May 2004.
[89] For a discussion of the level of convergence and the potential need for a legally binding instrument at international level see Ruiz Cairó, *The promotion of public health in EU external relations*, above n 48, at 236–38.

C. Conclusion

The WHO may be seen as a textbook example where the EU is welcome as an important 'payer' but does not amount to a real 'player'. While the 'REIO' treatment has overcome obstacles for the EU when negotiating international legal instruments or participating in Drafting Groups, the general status as observer still presents a handicap for effective participation by the EU Delegation, instead of the rotating Presidency, in the everyday work of the WHO Governing Bodies.

In order to overcome this shortcoming for EU visibility and continuity, a new attempt of upgrading the EU status is on its way in close collaboration with the Member States. Such an increased role in the WHO would also fuel the EU's ambition to shape a 'global health policy'. In this respect, the full participation in the negotiation of the pandemic treaty and the rights granted by the UNGA resolution of May 2011 could serve as a reference point how a future active participant status could look like.

IV. The EU at the International Labour Organization

As in the case of public health, the horizontal clause enshrined in Article 9 TFEU requires the Union to define and implement its policies with the objective to promote a high level of employment, the guarantee of adequate social protection and the fight against social exclusion. To this effect, internal market legislation can be enacted under Article 114 TFEU. The EU is also competent to enact anti-discrimination laws on the basis of Article 19 TFEU and may lay down rules for the protection of migrant workers (Articles 45–48 TFEU). In addition, Titles IX and X of the Treaty establish a complex system of EU competences in the area of employment and social policy. Importantly, some matters, such as pay, the right to association, the right to strike or the right to impose lock-outs, are explicitly excluded from the realm of EU powers (Article 153(5) TFEU). Accordingly, the powers on employment and social policy are generally shared between the EU and its Member States (see Article 4(2)(a)(b) TFEU), which triggers the same pre-emptive effect on Member States powers as described before in Chapters two and three for the EU's other policies. Finally, the EU Fundamental Rights Charter contains a number of social and economic rights in its Title IV on 'Solidarity'. They must be taken into account when the EU takes action, but do not confer new powers on the Union (Article 51(2) Fundamental Rights Charter).

A. Legal Status of the European Union

(i) General Status

The ILO is composed of states (Article 1(2) and 1(3) ILO Constitution), but organised in a tripartite way. Each state is represented by delegates from the government, and employers' and workers' associations. Under Article 12(1) of the ILO Constitution, the organisation can also cooperate with other international organisations and make 'appropriate arrangements' for them to participate in the work of the organisation without a right to vote (Article 12(2) ILO Convention). On that basis, the first President of the High Authority of the European Coal and Steel Community, Jean Monnet, signed a cooperation agreement with the ILO back in 1953,[90] and the European Economic Community followed suit in 1958 with an exchange of letters.[91] It was replaced first in 1989 and then in 2001 by a new version. Importantly, the ILO accepted in that series of letters that 'the Community, represented by the Commission, shall attend the meetings of the International Labour Conference and the Governing Body, to which it shall be regularly invited.'[92]

Against that background, the ECJ concluded in 1993 that the Community had acquired observer status in the ILO.[93] Accordingly, it can participate without a vote in the discussion in the International Labour Conference (ILC)[94] and all its technical committees.[95] The same is true also for the ILO Governing Body,[96] where the EU is admitted as observer next to the three permanent EU members (France, Germany and Italy) and three elected EU Member States. In contrast, the EU does not appear as a regional government group, or subgroup, at the ILO.[97]

Importantly, the Treaty of Lisbon had a significant impact on the way the EU made use of its observer status. Since 2013, internal coordination on ILO matters

[90] Agreement concerning Co-operation between the International Labour Organization and the European Coal and Steel Community, Official Bulletin of the ILO; Vol XXXVI, 1953, No 7.

[91] Exchange of letters of 7 July 1958, OJ 27, 27.4.1959, p 521.

[92] Exchange of letters of 21.12.1989 and 22.12.1989 between the European Commission and the International Labour Organization, OJ C 24, 1.2.1990, p 8 (at p 9), renewed by an exchange of letters of 14 May 2001, OJ C 165, 8.6.2001, p 23.

[93] ECJ, *Opinion 2/91* (Convention N° 170 of the ILO), of 19 March 1993, [1993] ECR I-1959, p 1066.

[94] See Art 2(3)(b) and Art 14(9) of the Standing Orders of the Conference.

[95] G Pons-Deladière, 'European Union Participation and Cooperation in ILO Institutions and Activities: An ILO Perspective', in C Kaddous (ed). *The European Union in International Organisations and Governance: Recent Developments* (Oxford, Hart Publishing, 2015), 93, 97–98.

[96] See Article 1.9. of the Standing Orders of the Governing Body.

[97] E Pichot and R Delarue, 'International Labour Organisation (ILO) – A dynamic and result-oriented cooperation with the EU and its Member States', in R Wessel and J Odermatt (eds), *Research Handbook on the European Union and International Organisations* (Cheltenham, Edward Elgar, 2019), 101, 103–4 with a description how these groups function in ILO practice.

has taken place together with social partners annually in the Social Questions Working Party in Brussels, complemented by on-the-spot coordination in Geneva organised by the EU Delegation for all ILO agenda points.[98] While the Commission expressed the position of the EU in a limited number of cases as an observer prior to 2009, the EU Delegation has assumed a greater role thereafter. Nowadays, there is a more structured division between the rotating Presidency and the EU Delegation to deliver Statements in various ILO bodies either 'on behalf of the EU' or 'on behalf of the EU and its Member States'.[99] The choice of the formula follows the division of competence and the result reached during intra-EU coordination in Geneva between the EU Delegation and the Member States. While the EU generally speaks through its Member States, EU actors will speak from behind the EU nameplate at political level (in the ILC plenary) or during informal and open sessions.[100] In no case, the EU may vote, but as the ILO usually strives for consensus voting is uncommon (except on programming, budget and staff issues).[101]

(ii) ILO Conventions

Under Article 19(2) the ILO Conference can adopt ILO Conventions by a two-thirds majority. They become binding on those states which are ratifying it. However, due to Article 19(5) of the 1919 ILO Constitution, the EU cannot become a formal party to any ILO Convention. Whenever such an ILO Convention falls fully or partially within EU competence, it has to be exercised through the medium of the Member States acting jointly in the Community's interest.[102] In the aftermath of the ECJ's Opinion, the Commission first issued recommendations to the Member States on the ratification of ILO Conventions (1996–2000).[103] However, it then changed course and proposed that the Council enact formal authorisation decisions, empowering the Member States to ratify ILO Conventions in the interests of the Union. To date, this has happened for five new Conventions[104]

[98] ibid, at 105.

[99] M Ferri, 'Coordination between the European Union and its Member States', in C Kaddous (ed), *The European Union in International Organisations and Governance: Recent Developments* (Oxford, Hart Publishing, 2015), 77, 91.

[100] Pichot and Delarue, 'International Labour Organisation (ILO)', above n 97, at 104.

[101] Pons-Deladière, 'European Union Participation and Cooperation in ILO Institutions and Activities: An ILO Perspective', above n 95, at 101.

[102] ECJ, *Opinion 2/91* (Convention N° 170 of the ILO), [1993] ECR I-1061, paras 36–38.

[103] Ferri, 'Coordination between the European Union and its Member States', above n 99, at 82 with further references.

[104] Council Decision 2005/367/EC, OJ L 136, 30.5.2005, p 1 (ILO Convention No 185); Council Decision 2007/431/EC, OJ L 161, 22.6.2007, p 63 (ILO Maritime Labour Convention, 2006); Council

and two Protocols.[105] A proposal for Member States to ratify the Violence and Harassment Convention 2019 (No 190) too, in the interest of the European Union, is pending.[106]

Conversely, when an 'old' ILO Convention falls foul of modern EU anti-discrimination law, the conflict must be resolved in favour of EU law. Hence their EU membership led almost all EU Members to terminate ILO Conventions No 45 and 89 on employment of women in underground work in mines or in night-work.[107]

B. Political Role of the European Union

Since the Treaty of Amsterdam, the EU has continuously raised its profile and started influencing the ILO agenda also at the political level.[108] The impact of the intensified cooperation between 2001 and 2011 was duly presented in an ILO brochure in 2012.[109]

As the standard-setting role of the ILO has reached a certain peak by having adopted over 190 Conventions and over 200 Recommendations, most energy is now being devoted to implementation. The ILO itself has adopted a couple of actions plans how to carry out, upon request, activities in ILO Member States. For example, the 1998 ILO Declaration on Fundamental Principles and Rights at Work leads to annual reports from Member States and a Global ILO Report. The 2008 ILO Declaration on Social Justice for a Fair Globalisation foresees several mechanisms for how the ILO can support its members, including through technical assistance and advisory services. Finally, the ILO runs a non-compliance system, where ILO members need to explain themselves in the Conference Committee on the Application of Standards and in the Governing Body.[110]

Decision 2010/321/EU, OJ L 145, 11.6.2010, p 12 (ILO Convention No 188); Council Decision 2014/51/EU, OJ L 32, 1.2.2014, p. 32 (ILO Convention No 189); Council Decision 2014/52/EU, OJ L 32, 1.2.2014, p 33 (ILO Convention No 170).

[105] Council Decision 2015/2037/EU, OJ L 298, 14.11.2015, p 23 (Protocol 2014 to the Forced Labour Convention of 1930); Council Decision 2015/2071/EU, OJ L 301, 18.11.2015, p 47 (Protocol of 2014 of the Forced Labour Convention as regards Articles 1 to 4 with regard to matters relating to judicial matters in criminal matters).

[106] COM(2020) 24 final of 22 January 2020.

[107] Ferri, 'Coordination between the European Union and its Member States', above n 99, at 83.

[108] R Delarue, 'ILO-EU cooperation on employment and financial issues', in J Wouters, F Hoffmeister and T Ruys (eds), *The United Nations and the European Union. An Ever Closer Partnership* (The Hague, TMC Asser Press, 2006), 93.

[109] ILO, 'The ILO and the EU, partners for decent work and social justice – Impact of ten years of cooperation' (ILO Office, 2012).

[110] Ferri, 'Coordination between the European Union and its Member States', above n 99, at 85.

One important recent example, where the EU was rather active in a country case, related to Quatar. Here, the EU's constant lobbying in the ILC and the Governing Body between 2014 and 2017 helped to improve the situation of vulnerable migrant workers.[111]

In the 'Common Agenda' of 2013, the EU and the ILO underlined their commitment to reinforce coordination especially when core labour standards are violated in third countries.[112] In this respect, the EU has put its powerful trade policy increasingly also at the service of ILO goals. Constitutionally, this development is covered by Article 207(1) second sentence TFEU, according to which the Union's trade policy should also pursue the general objectives of EU external action, as enshrined in Article 21(2) TEU. As the European Court of Justice held in the *Singapore Opinion*, EU trade policy is hence empowered (and even required) to pursue sustainable development goals as an integral part thereof.[113] In practice, the application of two instruments can illustrate this new trend: the EU Generalised Scheme of Preferences (GSP) Regulation and the EU trade and sustainability chapters in Free Trade Agreements.

(i) GSP Regulation

Under the Most-Favoured-Nation Clause of the WTO (Article I GATT), the EU is in principle bound to apply the same tariff to all its trading partners. One exception arises under the Enabling Clause, according to which any WTO Member may establish a generalised system of preferences for developing countries. Against that background, the EU allows preferential market access on a unilateral basis according to Regulation 978/2012.[114] Under this scheme, low- and middle-income countries benefit from certain preferences. On top of this, Article 9 grants even better access to countries which implement 27 international conventions related to labour and human rights, environmental and climate protection, and good governance. Famously, Annex VIII includes eight ILO Conventions, relating to the four core labour standards, which are all ratified by all EU Member States: prohibition of child labour (Conventions 138 and 182), prohibition of forced labour (Conventions 29 and 105); freedom of association and to bargain collectively

[111] Pichot and Delarue, 'International Labour Organisation (ILO) – A dynamic and result-oriented cooperation with the EU and its Member States', above n 97, at 106.

[112] Joint Conclusions of the 10th High Level Meeting between the International Labour Office and the European Commission, Brussels, 4–5 December 2012.

[113] ECJ, *Opinion 2/15* (Singapore FTA), ECLI:EU:C:376, paras 139–47.

[114] OJ L 303, 31.10.2012, p 1.

(Conventions 87 and 98), and non-discrimination (Conventions 100 and 111). Clearly, this special incentive scheme can be seen as an important EU contribution to the effective implementation of these ILO standards worldwide. In practice, this became very visible in the case of Myanmar, when the EU had to withdraw the incentives in 1997 because of the widespread use of forced labour by the military junta, but reinstated it upon democratic change in 2013.[115] This mirrored action at ILO level, which ended its measures against Myanmar to secure compliance with Convention No 29 (forced labour) under Article 33 of the ILO Constitution in the same year. Other EU GSP+ withdrawals concerned Belarus (2007), Sri Lanka (2010, reinstated 2017), but they were more related to human rights violations and democratic regression. In the case of Cambodia (2019), the Commission justified the withdrawal of preferences because of serious human *and* labour rights violations. The latter referred in particular to the curtailment of the right to association and collective bargaining.[116]

(ii) Trade and Sustainable Development Chapters

Another important connection between the EU and ILO Conventions exists in the EU chapters on trade and sustainable development (TSD) in EU free trade agreements (FTA). Since 2011, we can find them in any new EU FTA with the following partners: South Korea, Colombia/Peru/Ecuador, Central America, Ukraine, Georgia, Moldova, Canada, Japan, Singapore, Vietnam and the United Kingdom. Relevant chapters with Chile, China, Mercosur and Mexico await signature. The EU has also proposed provisions on TSD in its negotiations with Australia, Indonesia and New Zealand.

In substance, the TSD chapters, which usually reaffirm the rules agreed under the FTA in question (including investment protection rules), should not be used to engineer social or labour regression domestically. Moreover, they incorporate international conventions that the FTA partner has already subscribed to. By incorporating them into the FTA, the Parties make these standards a reference point for joint monitoring and submit them to a special dispute settlement procedure. Occasionally, as in the case of the EU–Vietnam FTA, a partner will also make

[115] For details see L Beke and N Hachez, 'The EU GSP: A preference for human rights and good governance? The Case of Myanmar', *Leuven Centre for Global Governance Studies, Working Paper No 155*, 2015, 18–25.

[116] European Commission Press Release of 11 February 2019, https://ec.europa.eu/commission/presscorner/detail/en/MEMO_19_988.

an effort to accede to new ILO Conventions during the process. Before that agreement was ratified, the Vietnamese government made important changes to the country's labour laws, ratified two out of three outstanding ILO core Conventions (Nos 98 and 105) and scheduled the ratification of the last one for 2023.[117]

The legal significance of TSD chapters is highly disputed.[118] It seems that a distinction between non-material and material breaches is warranted.[119] For non-material breaches, the FTA operates a specific dispute settlement procedure as *lex specialis* to the otherwise available instrument of a countermeasure.[120] They can hence be reviewed by a Panel, which issues a non-binding report, as occurred under the EU–Korea FTA. The reasoning of this Panel showed very convincingly that the implementation of core ILO standards has become a self-standing obligation under the trade agreement.[121] For material breaches, within the meaning of Article 60(2) of the Vienna Convention on the Law Treaties, it cannot be assumed that the Parties wished to exclude the possibility that the other side can resort to the suspension or even termination of the treaty. Such material breaches would normally occur when the breach of an ILO standard at the same time breaches fundamental rights protected by the essential elements clause of the FTA. Such a scenario would occur, for example, in the case of a violent crackdown on trade-unions, which would not only infringe the right to association, but also other civil and political rights, such as the right to life and liberty.

Importantly, in its June 2022 communication, the Commission announced to strengthen the enforcement of TSD commitments in future agreements even further. It wishes to

(a) extend the general state-to-state dispute settlement compliance stage to the TSD chapter;
(b) involve civil society bodies in monitoring the compliance stage; and

[117] K Marslev and C Staritz, 'Towards a stronger EU approach on the trade-labor nexus? The EU-Vietnam Free Trade Agreement, social struggles and labor reforms in Vietnam' (2023) 30(3) *Review of International Political Economy* 1125, https://www.tandfonline.com/doi/epdf/10.1080/09692290.2022.2056903?needAccess=true&role=button.

[118] For a good overview see M Bronckers and G Gruni, 'Retooling the Sustainability Standards in EU Free Trade Agreements' (2021) 24(1) *Journal of International Economic Law* 25.

[119] For an in depth-discussion see F Hoffmeister and A Siemer, 'The Legal Significance of Trade and Sustainability Chapters in the European Union's Free Trade Agreements', 2024(3) *Zeitschrift für Europarechtliche Studien* (forthcoming).

[120] ICJ, *Gabcikovo-Nagymaros Project (Hungary/Slovakia)*, Judgment, ICJ Reports 1997, p 7, at p 65, para 106.

[121] Panel of Experts established under the EU–Korea FTA, Report of 26 January 2021, paras 107–10.

(c) extend the possibility to apply trade sanctions in cases of failure to comply with obligations that materially defeats the object and purpose of the Paris Agreement on Climate Change or in serious instances of non-compliance with the ILO fundamental principles and rights at work.[122]

A few days later, on 30 June 2022, the negotiations for a trade agreement between the EU and New Zealand were concluded.[123] Signed on 9 July 2023, it 'is the first one to integrate the EU's new approach', according to the Commission.[124] Indeed, under Article 26.2 its TSD commitments are enforceable through the general dispute settlement mechanism. Moreover, the fulfilment of obligations clause (Article 27.4) contains in its paragraph 3 the following new language:

> (3) This Agreement forms part of the common institutional framework referred to in Article 52(1) of the Partnership Agreement. A Party may take appropriate measures relating to this Agreement in the event of a particularly serious and substantial violation of any of the obligations described in Article 2(1) or Article 8(1) of the Partnership Agreement as essential elements, which threatens international peace and security so as to require an immediate reaction. A Party may also take such appropriate measures relating to this Agreement in the event of an act or omission that materially defeats the object and purpose of the Paris Agreement. Those appropriate measures shall be taken in accordance with the procedure set out in Article 54 of the Partnership Agreement.

The clause hereby amends the 'essential elements clause' of the Partnership Agreement. Seen together, Articles 26.2 and 27.4(3) of the EU-NZ trade agreement create 'the possibility of trade sanctions as a matter of last resort, in instances of serious violations of core TSD commitments, namely the ILO fundamental principles and rights at work, and of the Paris Agreement on Climate Change'.[125]

C. Conclusion

The EU has publicly lent its efforts behind the ILO as a 'tried-and-trusted' organisation. Accepting the fact that the 1919 Constitution cannot be easily changed,

[122] European Commission, 'The power of trade partnerships: together for green and just economic growth', COM(2022) 409 final, p 12, Action point 19.
[123] Key elements of the EU-New Zealand trade agreement, https://policy.trade.ec.europa.eu/news/key-elements-eu-new-zealand-trade-agreement-2022-06-30_en.
[124] Press release, 'EU and New Zealand sign ambitious free trade agreement', https://ec.europa.eu/commission/presscorner/detail/en/ip_23_3715.
[125] Key elements of the EU-New Zealand trade agreement, n 122 above.

it is supporting the ILO as an active observer with political influence. It has also lent its powerful trade policy to encouraging third states enforcing ILO standards. This has taken the form of granting trade preferences under the GSP Regulation or the negotiation of an ever-stronger TSD chapter in EU FTAs with third countries. It may hence be argued that the EU is one of the best pupils in the ILO's class, although still sitting on the back-row.

V. Conclusion

Following the 2021 Communication from the High Representative and the European Commission on strengthening multilateralism, the EU has again focused more attention on its legal status and political role in international organisations. In the FAO, the membership status is largely satisfactory, but the EU internal practice of how to use it is not. Hence, it is high time to bring EU internal rules up to post-Lisbon standards and give EU actors in Rome a greater role compared to Brussels-driven processes in the hands of the rotating Presidency of the Council. In the WHO, the participation rights of the EU are hampered by its observer status in the Governing Bodies, which creates a dynamic to frequently use the rotating Presidency as EU spokesperson. Hence, a formal upgrade of the EU's legal status is necessary, conferring active participant rights by concluding a formal arrangement under Article 70 of the WHO Constitution. That would echo the successful status of the EU as an REIO in treaty negotiations under WHO auspices and correspond to the increasing weight of the Union in the aftermath of fighting the COVID-19 pandemic. Finally, the EU is also showing, vis-à-vis the ILO, how it can foster the objectives and activities of an organisation when it links its powerful trade tools with commitments by third countries to apply and enforce global ILO standards. Doing so, it can strengthen the idea that allowing the Union to play a bigger role in an international organisation will not only serve to sharpen the profile of the EU, but also help the UN specialised agency to fulfil its mandate more effectively.

In sum, in 2023 it is time for both the EU on the one hand, and the FAO, WHO and ILO on the other hand, not to 'rest and rust'.

6

The European Union and Global Economic Governance Institutions: Keeping the Flame Alive in Turbulent Times

JAN WOUTERS

I. Introduction

It is nowadays a truism to state that the multilateral global governance system that has been constructed over the past eight decades is facing an unprecedented number of challenges, if not an outright existential crisis. Numerous factors contribute to this. In the first place, one can point to a set of issues under the heading 'old age' ('*la viellesse*'). Most – though not all – of today's international institutions, especially those of the United Nations (UN) family, were created around the end of the Second World War, meaning they are currently in their eighth decade. Institutional reform has always proven to be difficult in global organisations, and that is definitely the case for the UN, the International Monetary Fund (IMF or Fund), and the World Bank. One may think of the long-standing need to reform the UN Security Council, which has operated since 1945 with its five permanent members from only the Northern hemisphere with veto rights that from time to time paralyse the decision-making process. Apart from this, some organisations suffer from stifling bureaucracy,[1] politicisation

[1] See in particular, regarding bureaucracy in the UN, T Weiss, *What's Wrong with the United Nations and How to Fix It*, 2nd edn (Cambridge, Polity, 2012), 114 ff; MW Bauer, C Knill and S Eckhard (eds), *International Bureaucracy. Challenges and Lessons for Public Administration Research* (London, Palgrave MacMillan, 2017).

and 'siloisation'.[2] While some emerging powers seek stronger representation and influence, and contest the legitimacy of current processes,[3] we also live in an era in which some of the 'established' powers – like the United States (US) – have become less predictable. But even a 'younger' international organisation like the World Trade Organization (WTO), established in 1995, is currently in a major crisis in light of the non-delivery of the Doha Development Agenda (DDA), the difficulty in agreeing on other new agreements, and the demise of its Appellate Body.

The European Union (EU or Union), itself in its seventh decade if one counts the earlier European Communities, is not immune to crises. While the EU has been able to reform itself through regular Treaty changes, it has, especially over the last 15 years, faced one crisis after another: the financial crisis, the Eurozone sovereign debt crisis, the migration crisis, Brexit, COVID-19, an energy crisis in light of Russia's military invasion of Ukraine, not to speak of the deterioration of the rule of law and of the quality of democracy in a number of its Member States. The acuteness of the situation has been well captured in the Global Strategy that High Representative for Foreign Affairs and Security Policy Federica Mogherini issued in June 2016:

> We live in times of existential crisis, within and beyond the European Union. Our Union is under threat. Our European project, which has brought unprecedented peace, prosperity and democracy, is being questioned. To the east, the European security order has been violated, while terrorism and violence plague North Africa and the Middle East, as well as Europe itself. Economic growth is yet to outpace demography in parts of Africa, security tensions in Asia are mounting, while climate change causes further disruption.[4]

Still, the EU has shown quite some resilience in addressing the aforementioned crises. The Union has also continued its commitment to 'a rules-based global order with multilateralism as its key principle and the United Nations at its core'.[5] In that light, the present chapter takes a closer look at the EU's role and position in some of the main institutions of global economic governance. This chapter

[2] For a case-study on 'siloisation', see C Anthes, *Institutional Roadblocks to Human Rights Mainstreaming in the FAO. A Tale of Silo Culture in the United Nations System* (Wiesbaden, Springer VS, 2020).

[3] On this issue see, inter alia, the contributions in M Egan, K Raube, J Wouters and J Chaisse (eds), *Contestation and Polarization in Global Governance. European Responses* (Cheltenham, Edward Elgar Publishing, 2023); E Johansson-Nogués, MC Vlaskamp and E Barbé (eds), *European Union Contested: Foreign Policy in a New Global Context* (Cham, Springer, 2020).

[4] European External Action Service, *Shared Vision, Common Action: A Stronger Europe. A Global Strategy for the European Union's Foreign and Security Policy* (Publications Office, June 2016), 13, https://www.eeas.europa.eu/sites/default/files/eugs_review_web_0.pdf.

[5] ibid, 8 and 15.

will ask whether the Union is keeping the flame of multilateralism alive in these turbulent times. It will consider both informal clubs, like the G7 and G20, and formal economic organisations, like the WTO, the Organisation for Economic Cooperation and Development (OECD), the 'Bretton Woods' organisations, ie the IMF and the World Bank Group, and UN economic fora.

We will examine first the informal clubs, ie the G7 and G20. The formal organisations are then studied in an order that has been deliberately chosen, namely from the WTO, where the EU enjoys full membership rights, onwards to the OECD, where the Union acts as a full participant, to UN economic fora, where the Union typically has to work from a more modest observer status, to the Bretton Woods institutions, where the EU presently has no status at all. Before turning to these respective institutions, it is worthwhile to develop some reflections on the EU's commitment to multilateralism and on the challenges that the Union faces in a world of international organisations that remains essentially 'Westphalian', ie based on the participation of sovereign nation states.[6]

II. The EU's Commitment to Multilateralism in a Westphalian World

The Union's commitment to multilateralism runs deep and finds its expression both in countless policy documents and statements – like the Global Strategy – and in its founding Treaties. As to the latter, Article 21(1) of the Treaty on European Union (TEU) emphatically provides that the Union 'shall promote multilateral solutions to common problems, in particular in the framework of the United Nations'. Just a little further in the same article, it is stated that the EU 'shall work for a high degree of cooperation in all fields of international relations, in order to … promote an international system based on stronger multilateral cooperation and good global governance'.[7] In terms of policy documents, as early as December 2003, when the European Council adopted the European Security Strategy,[8] strong emphasis was laid on the concept of 'effective multilateralism'.[9]

[6] For an earlier study in this respect, see J Wouters and A Chané, 'Brussels Meets Westphalia: The European Union and the United Nations', in P Eeckhout and M Lopez-Escudero (eds), *The European Union's External Action in Times of Crisis* (Oxford, Hart Publishing, 2016), 299–323.

[7] Article 21(2) first para and subpara (h), TEU.

[8] Brussels European Council, Presidency Conclusions, 12–13 December 2003, para 84.

[9] The concept of 'effective multilateralism' cannot be found anymore in the 2016 Global Strategy, although the latter refers at one juncture, when it addresses transatlantic cooperation, to 'effective global governance': Global Strategy, n 4 above, 36.

> The European Council reaffirms the deeply rooted commitment of the European Union to making effective multilateralism a central element of its external action, with at its heart a strong UN.[10]

The European Security Strategy itself states:

> In a world of global threats, global markets and global media, our security and prosperity increasingly depend on an effective multilateral system. The development of a stronger international society, well-functioning international institutions and a rule-based international order is our objective.[11]

The 2003 Strategy's strong focus on the EU's contribution to 'an effective multilateral system leading to a fairer, safer and more united world'[12] can be explained against the background of the '*annus horribilis*' that the year 2003 was for multilateralism: the US–UK invasion of Iraq had exposed great divisions within both the UN Security Council and the EU itself.[13] With a common analysis of both the threats to the Union's security and the methods and tools for addressing them, the European Security Strategy brought EU Member States back together.

It remains a fact, though, that the Union has to achieve its goal of stronger multilateral cooperation in a world that still favours nation states as the primary actors of international law and international relations. Most international organisations, especially those of the UN system, only allow states as members. One can even submit that this 'Westphalian' characteristic has again become stronger in today's international context, which faces not just a crisis of multilateralism,[14] but also increasing tensions and a renewed arms race between great powers and a resurgence of both non-international and international armed conflicts.

To this somewhat inhospitable international environment, one has to add some of the EU's internal challenges to making an effective contribution to global governance. These include the complexities of the delineation of competences between the Union and its Member States in international affairs and more

[10] Brussels European Council, Presidency Conclusions, 12–13 December 2003, para 91.

[11] European Council, *A Secure Europe in a Better World. European Security Strategy* (Publications Office, December 2003), 9.

[12] ibid, 14.

[13] See, inter alia, A Graf Lambsdorff, 'Nach dem "annus horribilis": Das Reformjahr 2005 muss die UN wieder zukunftsfest machen' (May 2005) *Internationale Politik* 114–20.

[14] See, inter alia, L Gruszczynski, MJ Menkes, V Bílková and PD Farah (eds), *The Crisis of Multilateral Legal Order: Causes, Dynamics and Implications* (New York, Routledge, 2023); A Guilbaud, F Petiteville and F Ramel (eds), *Crisis of Multilateralism? Challenges and Resilience* (Cham, Palgrave MacMillan, 2023).

generally the challenges of coordinating the foreign policies of 27 sovereign states. Part of the complexity has to do with the way in which the EU is represented in its relations with the wider world. While the Treaty of Lisbon has created a broadly consistent system of external representation, introducing 'permanent' interlocutors like the President of the European Council, the High Representative of the European Union for Foreign Affairs and Security Policy (who is also a vice-president of the European Commission) and the Union Delegations to represent the Union 'in third countries and at international organisations' (Article 221(1) Treaty on the Functioning of the European Union (TFEU)), the EU Member States continue to have a competitive edge in the global multilateral system in that they are full members of the main organisations, from the UN to the UN specialised agencies, including the Bretton Woods institutions. Only in the G20 and the WTO is the Union itself a full member.

The current international system does not only structurally impede the EU's full role. A great number of third countries are not keen on welcoming the Union as a strong actor in international affairs going beyond trade. The Global South – embodied in organisations such as the G77 and the Non-Aligned Movement (NAM) – retains the collective memory of Europe's colonial past, generating suspicion whenever the EU talks about promoting human rights and democracy. In the field of economic relations, the dominance of the EU is viewed with suspicion, too. A number of countries, in particular the expanding BRICS,[15] are strongly wedded to principles of sovereignty and non-intervention that are hard to reconcile with many of the Union's external policies. Regrettably, in 2020 the EU also lost the United Kingdom, which is not only a permanent member of the UN Security Council, but also one of Europe's two nuclear powers and one of the strongest contributors to international development aid. Russia has been emphatically called a foe by Europe's politicians since its large-scale military aggression against Ukraine in 2022. And the Union itself calls China today a cooperation partner or negotiation partner, as well as an economic competitor and 'a systemic rival promoting alternative models of governance'.[16]

[15] At the 15th BRICS Summit in Sandton, South Africa on 22–24 August 2023, it was decided to expand the membership of BRICS (Brazil, Russia, India, China, South Africa) from 1 January 2024 to include Argentina, Egypt, Ethiopia, Iran, Saudi Arabia, and the United Arab Emirates: *Johannesburg II Declaration. BRICS and Africa: Partnership for Mutually Accelerated Growth, Sustainable Development and Inclusive Multilateralism*, 23 August 2023, para 91.

[16] European Commission and High Representative of the Union for Foreign Affairs and Security Policy, Joint Communication to the European Parliament, the European Council and the Council, *EU-China – A Strategic Outlook*, JOIN/2019/5 final, 12 April 2019, 1.

III. The Informal Clubs: The EU at the G7 and G20[17]

A. The EU at the G7

The Group of Seven (G7) was established initially as the G6 in Rambouillet in 1975 by the major industrialised countries: the US, the United Kingdom, France, Germany, Japan and Italy.[18] Soon after, they were joined by Canada and the then European Economic Community (EEC). In 1998, Russia was admitted to the G7, except for certain financial and economic affairs, forming the G8. However, Russia's membership was suspended in the Spring of 2014 following its annexation of Crimea.[19] Originally, the G7 was an exclusively economic club. Its genesis is specifically connected with the collapse of the Bretton Woods system of fixed exchange rates, the oil crisis, and the subsequent economic recession in the 1970s. Later the G7's political mission, the spreading of the 'market democracy' model, was strongly emphasised. After the fall of the Berlin Wall, attention shifted to the post-communist transition, the reform of international institutions, and global issues such as democratisation, the environment, drugs, and terrorism. The G7 operates on the basis of annual summits of the Heads of State and Government of its members. Around these meetings, a whole system of ministerial meetings has grown to discuss and build consensus on specific challenges in various domains. The G7 ministers of different subject-matter fields have met as well on a more or less regular basis. The G7 ministers of trade, finance, and foreign affairs meet at least annually. For other issue areas, meetings take place more irregularly. The meetings are chaired by the annual rotating Presidency, which hosts the summit and preparatory meetings and manages the agenda. Moreover, the coordination and planning among the 'sherpas', ie the personal representatives of the Heads of State or Government, also takes place in the run-up to each summit. In addition, there are also ad hoc meetings and ad hoc working groups. The lack of institutionalisation of the G7/G8 has at the same time proven to be the strength of the

[17] This section builds on J Wouters, C Ryngaert, T Ruys and G De Baere, *International Law: a European Perspective* (Oxford, Hart Publishing, 2018), 369–71; J Wouters and S Van Kerckhoven, 'The Role of the EU in the G7 in Times of Populism, Nationalism and Anti-Globalism', in C Oldani and J Wouters (eds), *The G7, Antiglobalism and the Governance of Globalization* (London, Routledge, 2018), 68–89; J Wouters, S Van Kerckhoven and J Odermatt, 'The EU at the G20 and the G20's Impact on the EU', in B Van Vooren, S Blockmans and J Wouters (eds), *The EU's Role in Global Governance: The Legal Dimension* (Oxford, Oxford University Press, 2013), 259–71.

[18] H Dobson, *The Group of 7/8* (Abington, Routledge, 2007), 5.

[19] See, inter alia, A Kortunov, 'Moscou et le G7: un drame en trois actes, prologue, et épilogue' (2019) (2) *Politique étrangère* 75–89.

forum. Informal meetings enable a faster reaching of consensus than via formalised multilateral channels.

As far as the EU's participation in the G7 is concerned, it can be noted that, in addition to the EU Member States represented at the G7, the President of the European Commission has been invited since 1977.[20] The rotating Council Presidency has irregularly attended G7 summits from 1982 onwards. Despite being structurally disadvantaged at the outset, the Commission has managed to emancipate itself within the G7 over time. This relates both to the fact that European integration has continuously expanded and deepened since the Commission's first participation in the G7 summitry, and the fact that the latter provided the perfect opportunity for the Commission to increase its international relevance.[21] For matters like trade policy and development aid, the Commission's contribution has proved vital.

In the post-Lisbon era, the EU is represented at G7 summits by both the President of the Commission and the President of the European Council. At G7 meetings of ministers of foreign affairs, the Union is represented by the High Representative. For example, High Representative Josep Borrell participated in the G7 Foreign Ministers' meeting in Tokyo in November 2023.[22] Moreover, the legal advisors of the G7 Foreign Ministers have held regular informal exchanges since March 2022 to give legal input for the ever-increasing number of G7 Foreign Ministers or Leaders' Statements. For the EU, the Legal Department of the European External Action Service is in charge of this network.

At ministerial meetings on other issues, a Commissioner will typically participate. Thus, European Commission Executive Vice-President and Commissioner for Trade, Valdis Dombrovskis, participated in the G7 Trade Ministers' meeting in Osaka-Sakai on 29 October 2023.[23] At services level, DG TRADE and DG FISMA[24] are in charge of coordinating the Commission positions in these fields.

[20] This was the third G7 summit in London of 7–8 May 1977, with Commission President Roy Jenkins attending; the UK then held the rotating Council Presidency. Initially this happened against the will of the progenitor of the G7 process, the French President Valéry Giscard d'Estaing: Dobson, n 18 above, 57; S Hainsworth, 'Coming of Age: The European Community and The Economic Summit', *G7 Research Group*, Country Study no 7, www.g7.utoronto.ca/scholar/hainsworth1990/index.html.

[21] See A Niemann and J Huigens, 'The European Union's role in the G8: a principal-agent perspective' (2011) 18(3) *Journal of European Public Policy* 420.

[22] See https://www.eeas.europa.eu/eeas/g7-foreign-ministers-meeting-press-remarks-high-representativevice-president-josep-borrell-meeting_en?s=169.

[23] On the margins of the G7 meeting, the EU and Japan concluded an agreement on cross-border data flows at their EU-Japan High-Level Economic Dialogue (HLED) on 28 October 2023, https://ec.europa.eu/commission/presscorner/detail/en/IP_23_5378.

[24] EU Directorates-General for Trade and for Financial Stability, Financial Services and Capital Markets Union, respectively.

The EU is the only G7 actor with both direct and indirect participation in the Group: directly through the participation of the President of the European Council and President of the European Commission in summits and of the rotating Presidency and Commissioners in ministerial meetings; and indirectly through the EU Member States that are G7 members in their own right, ie France, Germany, and Italy. This unique characteristic is partly a result of the concern that EU Member States which are not in and of themselves members of the G7 felt treated as second-class countries in G7 proceedings. As they were not represented, they lacked a voice in discussing global challenges and their solutions. For that reason, excluded Member States advocated for some form of representation of the then Community. They were joined in this endeavour by the Commission itself. The Commission lobbied intensively for representation at the summits based on the exclusive competence of the Community in certain policy fields, such as international trade.[25] In this context, it deemed an agreement on export credits at the 1976 Puerto Rico G7 summit illegitimate because it interfered with Community competences.[26] In 1987, the Commission for the first time took the lead in preparing a G7 working paper dealing with the problem of sub-Saharan African debt, which served as the basis for summit discussions.[27] A few years later, the G7 tasked the Commission with the coordination of Western aid to the former Eastern Bloc countries, in the aftermath of the collapse of communism.[28] The resulting PHARE programme, which was decided upon at the G7's Paris Summit, clearly demonstrated the recognition of the Commission's role in the G7. Compared to the Commission's involvement, the rotating Council Presidency failed to have an impact until the establishment of a permanent President of the European Council by the Treaty of Lisbon.

Coordination prior to G7 meetings between EU Member States and the EU is not extensive.[29] According to the Commission, a legal mandate is not necessary for the EU to participate in the G7's work, since no formal decision-making takes place at the G7 and discussions remain mostly political.[30] Informative briefings from the

[25] See P Debaere, 'The EU's role and performance within the G20 in the area of finance and development', in Joachim A Koops and Gjovalin Macaj (eds), *The European Union as a Diplomatic Actor* (Basingstoke, Palgrave Macmillan, 2015), 120 ff.
[26] Niemann and Huigens, n 21 above, 424.
[27] Hainsworth, n 20 above.
[28] Dobson, n 18 above, 57.
[29] P Debaere and J Orbie, 'The European Union in the Gx system', in KE Jørgensen and KV Laatikainen (eds), *Routledge Handbook on the European Union and International Institutions: Performance, Policy, Power* (London, Routledge, 2012), 311, at 318.
[30] J Huigens and A Niemann, 'The G8 ½: the EU's Contested and Ambiguous Actorness in the G8' (2011) 24(4) *Cambridge Review of International Affairs* 629, at 640.

Commission to the EU's Committee of Permanent Representatives (COREPER) are conducted before and after every G7 summit. Additionally, the European Council – who in the past often met before a G7 summit[31] – can adopt a coordinated position. Coordination is stronger for financial and monetary matters, with a division of work between the Eurozone Group, focusing on monetary issues, and the Commission, dealing with more general economic issues.[32] Sometimes G7 matters are also discussed during meetings of the EU's Economic and Financial Affairs (ECOFIN) Council.[33]

It has been argued in earlier scholarship that EU–G7 cooperation is a win-win. The relationship between the EU and the G7 is interdependent and mutually reinforcing. The EU helps to improve the effectiveness of the G7, while participation in the meetings of the latter reinforces the international legitimacy of the EU.[34]

While the EU is not a member of the G7, it is definitely an 'active participant' and is sometimes even referred to as the Group's 'eighth member'.[35] The EU was given the unique opportunity to host for the first time a G7 Summit in Brussels in 2014 after Russia was suspended from the G8.[36]

B. The EU at the G20

The Group of Twenty (G20) is an informal forum that brings together the 20 systemically most important economies worldwide. It comprises 19 countries and the EU; the 2023 New Delhi Summit welcomed the African Union as a member.[37]

[31] See, with examples, Hainsworth, n 20 above.

[32] See in more detail S Nasra, D Lesage, J Orbie, T Van de Graaf and M Vermeiren, 'The EU in the G8 system: Assessing EU member states' involvement', *EUI Working Papers* 2009, RCAS 2009/45, 5–6.

[33] See eg the ECOFIN Council of 18 January 2022, where the German delegation updated ministers on the priorities of the G7 Presidency, https://www.consilium.europa.eu/en/meetings/ecofin/2022/01/18/.

[34] See M Larionova and M Rakhmangulov, 'The EU as a Global Actor: Priorities, Functions, Values', in M Larionova (ed), *The EU in the G8: Promoting Consensus and Concerted Actions for Global Public Goods* (London, Routledge, 2012), 285 ff.

[35] The Commission on one of its websites refers to the EU as a 'non-enumerated member' of the G7, https://agriculture.ec.europa.eu/international/international-cooperation/international-organisations/g7_en.

[36] See P Debaere, 'The Brussels Summit: Implications for the EU's Role in the G7' (2017) 68(3) *Studia Diplomatica* 109.

[37] See 'G20 New Delhi Leaders' Declaration', New Delhi, India, 9–10 September 2023, para 76, https://www.g20.org/content/dam/gtwenty/gtwenty_new/document/G20-New-Delhi-Leaders-Declaration.pdf. The EU supported this decision: in its conclusions of 29–30 June 2023, para 40, the European Council stated: 'In line with commitments on multilateralism made at the European Union–African Union Summit of 17–18 February 2022, the European Council supports the African Union's reinforced presence in international forums, notably in the G20.' The African Union is a continental union, formed in 2002 with 55 Member States, which replaced the Organisation of African Unity.

The G20 was first conceived in reaction to the G7's perceived failure to adequately respond to the financial crisis that hit East Asia in the late 1990s. The G20 Finance Ministers and Central Bank Governors convened for the first time in Berlin on 15–16 December 1999.[38] G20 meetings were lifted to the level of Heads of State and Government with the Washington Summit in 2008, in response to the (Western) financial crisis. Since then, a G20 Leaders' Summit has been held annually (initially, during the crisis, twice a year), whereas the G20 Finance Ministers and Central Bank Governors meet twice a year. Since the 2009 Pittsburg Summit, the G20 has called itself 'the premier forum for our international economic cooperation'.[39]

Depending on the topics under debate, the G20 Chair may also invite other states to attend the G20 Leaders' Summits. In addition, a number of heads of international institutions (IMF, World Bank, WTO, ILO, OECD) are usually also represented at G20 Summits. In economic terms, the G20 membership represents five continents, 66 per cent of the world's population, approximately 85 per cent of global GDP and approximately 75 per cent of international trade.[40]

Like the G7, the G20's activities are not governed by an underlying founding document or charter. The G20 does not impose any formal membership conditions, nor does it employ any formal rules or mechanisms with respect to voting, decision-making or dispute resolution. It also lacks a permanent secretariat and official headquarters. It is therefore not an international organisation.[41] At its core, the G20 is designed as an international forum for debate and consensus-seeking on all matters related to global economic governance. As such, it is an inherently informal network. The informality that characterises the G20 also applies to its decision-making, as the outcome does not amount to a legally binding commitment for its members. The G20 has taken great care to avoid any institutionalisation of its structure. The final declaration of the 2011 G20 Summit in Cannes, for example, states that the G20 'is a leader-led and informal group and it should remain so'.[42] The Presidency of the G20 is held by the members on a rotating basis, and the Chair is selected annually from a different regional group (of which there

[38] PI Hajnal, *The G20. Evolution, Interrelationships, Documentation*, 2nd edn (New York, Routledge, 2019), 15.

[39] G20 Leaders' Statement. The Pittsburgh Summit, 24–25 September 2009, paras 19 and 50, https://www.oecd.org/g20/summits/pittsburgh/G20-Pittsburgh-Leaders-Declaration.pdf.

[40] See, inter alia, https://www.g20.org/en/about-the-g20/e-book-brasil-na-presidencia-do-g20, at 4.

[41] See for a more detailed analysis PH Henley and N Blokker, 'The Group of 20: A Short Legal Anatomy from the Perspective of International Institutional Law' (2013) 14(2) *Melbourne Journal of International Law* 550.

[42] Cannes Summit Final Declaration, 'Building our Common Future: Renewed Collective Action for the Benefit of All', 4 November 2011, para 91, http://www.g20.utoronto.ca/2011/2011-cannes-declaration-111104-en.html.

are five). In order to ensure continuity, management of the G20 is structured in a so-called 'Troika', which is composed of the preceding, current, and subsequent Chairs. There is strong cooperation between the G20 and other – formal – international organisations, such as the IMF, WTO and OECD. The G20 also works in cooperation with the Financial Stability Board (FSB), which it created at its London Summit in April 2009, and the Basel Committee on Banking Supervision.[43]

The EU is very active in the G20. Like in the G7, European interests are represented both directly through the EU's membership in the G20 and indirectly through the membership of France, Germany, and Italy. Spain and the Netherlands, while not members, are invited to G20 meetings. At the leaders' level, both the President of the European Commission and the President of the European Council attend the G20 summits. For the meeting of Finance Ministers and Central Bank Governors, the Commissioner for Economic and Monetary Affairs, the rotating Council Presidency, and the President of the European Central Bank (ECB) represent the EU.

In order to give a voice to the EU Member States that are not members of the G20, the European Council used to adopt 'Agreed Language' in preparation for G20 summits. Nowadays, references to G20 summits are scarce in European Council conclusions.[44] During the terms of President of the European Council Donald Tusk and President of the European Commission Jean-Claude Juncker, they jointly sent preparatory letters to the other Member States before G20 summits.[45]

[43] On the EU's relationship with the FSB and the Basel Committee on Banking Supervision, see S Donnelly, 'Financial Stability Board (FSB), Bank for International Settlements (BIS) and Financial Market Regulation Bodies. ECB and Commission Participation alongside the Member States', in RA Wessel and J Odermatt (eds), *Research Handbook on the European Union and International Organizations* (Cheltenham, Edward Elgar Publishing, 2019), 360–85.

[44] Two relatively rare recent references can be found in the European Council's conclusions of 24–25 June 2021, para 10: 'The European Council looks forward to swift progress on global corporate tax reform on a consensual basis within the G20/OECD framework'; and in its conclusions of 21–22 October 2021, para 5: 'In the context of the upcoming G20 meeting and in view of the special session of the World Health Assembly in November, the European Council underlines its support for a strong, central role of the World Health Organization in future global health governance and for the objective of agreeing an international treaty on pandemics.'

[45] See eg, before the G20 summit in Hamburg in 2017, their joint letter of 4 July 2017, https://www.consilium.europa.eu/media/23956/letter-to-ms-on-g20-summit-in-hamburg.pdf; before the G20 summit in Argentina in 2018, their joint letter of 27 November 2018, https://www.consilium.europa.eu/media/37182/joint_letter_of_presidents_donald_tusk_and_jean-claude_juncker_on_the_upcoming_g20_summit_0.pdf; and, before the Osaka Summit in 2019, their joint letter of 26 June 2019, https://www.consilium.europa.eu/media/40040/190626-letter-to-member-states-on-g20-summit-in-osaka.pdf. In the previous political cycle, Commission President Barroso and European Council President Van Rompuy had also sent joint letters to the European Council members: see eg joint letter on the G20 summit in Cannes, 7 October 2011, https://www.consilium.europa.eu/media/26582/125007.pdf. They also wrote joint letters to the other G20 members in preparation for, respectively, the Toronto Summit (joint letter of 22 June 2010) and the Seoul Summit (joint letter of 5 November 2010).

In ECOFIN, the preparation of G20 meetings of Finance Ministers and Central Bank Governors, as well as the feedback from past meetings, are discussed.[46]

The EU has been fairly successful in setting the G20 agenda and in implementing G20 decisions. However, it has regularly failed to get specific objectives adopted – such as, in 2010–11, a global tax on financial institutions – and has also witnessed some issues being included in the agenda that were not part of the European agenda, such as currency wars and global financial safety nets.[47] In recent times, the Union has voiced its frustration at the lack of progress by the G20 on climate issues.[48] It is often said that, after the successful way in which the G20 managed to bring the financial crisis under control in 2008–10, it has not been able to muster similar political will and consensus to address wider issues, including climate change, the COVID-19 pandemic,[49] and the recent Russia–Ukraine war.[50]

IV. Full Actorness: The EU at the WTO

The EU's position and role in the WTO can be seen in many respects as the 'quasi-ideal' situation for the Union in terms of engagement with global economic governance institutions. The EU is indeed a co-founding member of the WTO and has full membership in the organisation, alongside the individual membership of the 27 EU Member States (Article XI of the Marrakesh Agreement). Essentially, the Union is represented in the WTO by the Commission. It is the EU Trade Commissioner who represents the EU in the WTO's highest decision-making body, ie the Ministerial Conference. The Commission also represents the EU in

[46] See, for instance, ECOFIN Council, 14 March 2023, https://www.consilium.europa.eu/en/meetings/ecofin/2023/03/14/, where the Presidency and the Commission informed the Council of the main outcomes of the G20 meeting of Finance Ministers and Central Bank Governors held on 24–25 February 2023, and where the Council gave guidance for further work regarding the preparations for the upcoming G20 meeting of Finance Ministers and Central Bank Governors.

[47] Wouters, Van Kerckhoven and Odermatt, n 17, at 266.

[48] See the explicit criticism by Environment Commissioner Virginijus Sinkevičius after a G20 ministerial meeting on climate change of 28 July 2023: 'At the end of our meeting today, is the glass half full or half empty? … It is certainly empty when we look at where we stand on G20 commitments to address climate change – we simply are nowhere': https://www.politico.eu/article/we-simply-are-nowhere-eu-slams-lack-of-progress-at-g20-climate-meeting/. See also P Paramasivam and K Abnett, 'G20 nations fail to agree on emission reduction targets at talks', 28 July 2023, https://www.reuters.com/sustainability/no-g20-consensus-emissions-mitigation-targets-talks-source-2023-07-28/.

[49] For a critical appraisal, see M Ruelens and J Wouters, 'Covid-19 and the Trials and Tribulations of Global Health Governance' (2021) 16(2) *International Organisations Research Journal* 70, at 76–79.

[50] See 'G20 fails to reach consensus on Russia-Ukraine war', 25 February 2023, https://www.euractiv.com/section/global-europe/news/g20-fails-to-reach-consensus-on-russia-ukraine-war/.

the WTO's General Council, the sectoral councils, and the subsidiary bodies of the WTO.[51] Interestingly, in nearly all WTO meetings it is only the Commission representative that speaks on behalf of the EU. Only in the Budgetary Committee does the EU not participate and EU Member States intervene themselves. Importantly, even on staff matters, there is increased EU coordination.[52]

The representation of the EU by the Commission in the WTO is obviously linked to the exclusive competence of the EU for the Common Commercial Policy (Article 3(1)(e) TFEU). The WTO Agreement itself has an interesting provision about the number of votes at the disposal of the Union: Article IX(1) stipulates that, '[w]here the European Communities exercise their right to vote, they shall have a number of votes equal to the number of their member States which are Members of the WTO'. However, in practice no voting occurs, as the WTO operates by consensus.[53]

The EU is also a very prominent player in the WTO Dispute Settlement System. Since 1995, the Union, through the legal service of the Commission, has been involved in more than 200 cases as complainant or defendant. In more than 200 other cases it has requested third-party status.[54]

The fate of the WTO, which today is in an existential crisis because of the crisis of the Appellate Body, the non-delivery on the DDA, and the sheer difficulty of reaching consensus on any new agreements,[55] offers the opportunity to critically consider the EU's commitment to multilateralism in the area of international trade. After a very disappointing Cancún Ministerial Conference in 2003, the European Commission redefined the priorities of EU trade policy with its 2006 communication *Global Europe: Competing in the World*.[56] Henceforth, while it

[51] For the overview of WTO organs, see https://www.wto.org/english/tratop_e/monitor_e/monitor_e.htm#councils_committees.

[52] See F Hoffmeister, 'The European Union in the World Trade Organisation – A Model for the EU's Status in International Organisations?', in C Kaddous (ed), *The European Union in International Organisations and Global Governance: Recent Developments* (Oxford, Hart Publishing, 2015), 121 at 133–34.

[53] This explains why decision-making is so painstakingly slow. For a critical and alternative approach, see AB Zampetti, P Low and PC Mavroidis, 'Consensus Decision-Making and Legislative Inertia at the WTO: Can International Law Help?' (2022) 56(1) *Journal of World Trade* 1.

[54] See https://policy.trade.ec.europa.eu/enforcement-and-protection/dispute-settlement/wto-dispute-settlement/wto-disputes-cases-involving-eu_en; see, inter alia, GM Durán, 'The EU and its Member States in WTO Dispute Settlement: A "Competence Model" or a Case Apart for Managing International Responsibility?' in M Cremona, A Thies and RA Wessel (eds), *The European Union and International Dispute Settlement* (Oxford, Hart Publishing, 2017), 237.

[55] See, inter alia, B Hoekman and PC Mavroidis, 'WTO Reform: Back to the Past to Build for the Future' (2021) 12(S3) *Global Policy* 5.

[56] European Commission, Communication to the Council, the European Parliament, the European Economic and Social Committee and the Committee of the Regions, *Global Europe: Competing in the World*, COM(2006) 567 final, 4 October 2006.

remained committed to the WTO and the DDA, the Union would prioritise deep and preferential trade agreements with its strategic partners. Indeed, according to the Commission:

> Free Trade Agreements (FTAs), if approached with care, can build on WTO and other international rules by going further and faster in promoting openness and integration, by tackling issues which are not ready for multilateral discussion and by preparing the ground for the next level of multilateral liberalisation. Many key issues, including investment, public procurement, competition, other regulatory issues and IPR enforcement, which remain outside the WTO at this time can be addressed through FTAs.[57]

This new approach has resulted in an enormous flurry of new, deep, and comprehensive preferential trade agreements with a great number of countries and partners, from Korea, Canada, Singapore, and Japan to Chile, Mexico, New Zealand, and Mercosur,[58] with many more in the pipeline.[59]

Still, it would be wrong to conclude that the EU has given up on the WTO. When the Appellate Body ceased to function in December 2019 (because the US blocked (re-) appointments of members), the Union took the lead alongside a number of partners in proposing a temporary backstop mechanism, the Multiparty Interim Appeal Arbitration Arrangement (MPIA), to which 53 WTO Members have adhered in the meantime.[60] The European Commission has also made various contributions to the debate on WTO reform.[61] But the EU has also recently given itself stronger unilateral enforcement mechanisms with its recently amended Enforcement Regulation, enabling a swift suspension of obligations under trade agreements while the dispute settlement mechanism is blocked.[62]

[57] ibid, at 8.

[58] Mercosur is a South American trade bloc with four full members (Argentina, Brazil, Uruguay and Paraguay), and seven associate countries (Bolivia, Chile, Colombia, Ecuador, Guyana, Peru and Suriname).

[59] For a full overview, with information on the current state of negotiations, adoption or ratification, see https://policy.trade.ec.europa.eu/eu-trade-relationships-country-and-region/negotiations-and-agreements_en. A different but very informative overview can also be found on the website of the Council: https://www.consilium.europa.eu/en/policies/trade-policy/trade-agreements/.

[60] See, inter alia, N Gervais and A-Ph Ouellet, 'Echapper belle: AMPA ou MPAI. Quatre lettres au secours du système de règlement des différends à l'OMC' (2019) 32(2) *Revue québécoise de droit international* 29; J Pauwelyn, 'The WTO's Multi-Party Interim Appeal Arbitration Arrangement (MPIA): What's New?' (2023) *World Trade Review* 1; O Starshinova, 'Is the MPIA a Solution to the WTO Appellate Body Crisis?' (2021) 55(5) *Journal of World Trade* 787.

[61] See for a critical discussion J Wouters and V Hegde, 'Reform of Global Trade Governance: the Role of the European Union' (2022) 44(5) *Journal of European Integration* 715.

[62] Regulation (EU) 2021/167 of the European Parliament and of the Council of 10 February 2021 amending Regulation (EU) No 654/2014 concerning the exercise of the Union's rights for the application and enforcement of international trade rules. See F Graafsma, 'The Revised Enforcement Regulation (No 2021/167) and Some Possible Effects on EU Trade Disputes' (2022) 17 (7/8) *Global Trade and Customs Journal* 289.

V. Full Participant: The EU at the OECD

The OECD is a fascinating and dynamic organisation, established in 1960 in Paris.[63] It currently boasts 38 Member Countries, of which 22 are EU Member States. But it is definitely no longer a primarily Western institution: apart from Japan, Mexico, Chile and Israel, Colombia and Costa Rica are among the most recent accessions.[64] The OECD describes itself as 'a community of nations committed to the values of democracy based on rule of law and human rights, and adherence to open and transparent market-economy principles.'[65] The OECD aims to maximise the promotion of economic growth, employment, and living standards in a stable financial climate, as well as further liberalise world trade. To achieve these aims, its activities focus on information exchange, consultations, and conducting research studies. The OECD bodies are the Council, the Executive Committee, and the Secretariat, headed by a Secretary General. As the decision-making body of the OECD, the Council consists of one representative of each Member Country, as well as one representative of the European Commission. It frequently gathers at the level of permanent representatives, but it also meets once a year on the ministerial level to discuss important issues and determine priorities. The Council, which operates based on consensus, may make binding decisions, issue recommendations, and adopt conventions. Under the supervision of the Council, around 250 specialised committees, working groups, and expert groups (subsidiary bodies within the meaning of Article 9 OECD Convention) operate, in which Member Countries develop ideas and evaluate progress in specific policy domains, such as trade, science, work, education, financial markets, and the economy.[66]

As indicated, it is the Commission that takes part, for the EU, in the work of the OECD, based on Supplementary Protocol No 1 to the OECD's 1960

[63] The OECD succeeded the Organisation for European Economic Co-operation (OEEC), which was established with 18 Western European countries in 1948 to administer the Marshall Plan: https://www.oecd.org/about/history/oeec/. On the European Communities' engagement with the OEEC and their participation in the negotiations for the establishment of the OECD, see F Vlastou-Dimopoulou, 'Organisation for Economic Co-operation and Development (OECD). The EU's Position as a De Facto Member', in Wessel and Odermatt (eds), *Research Handbook on the European Union and International Organizations*, n 43 above, 316, at 319–21.

[64] On 25 January 2022, the OECD Council decided to take the first step in accession discussions with six candidate countries – Argentina, Brazil, Bulgaria, Croatia, Peru and Romania, https://www.oecd.org/newsroom/oecd-takes-first-step-in-accession-discussions-with-argentina-brazil-bulgaria-croatia-peru-and-romania.htm.

[65] OECD 50th Anniversary Vision Statement, C/MIN(2011) 6, Meeting of the OECD Council at Ministerial Level, Paris, 25–26 May 2011: https://www.oecd.org/mcm/48064973.pdf.

[66] This paragraph is based on Wouters, Ryngaert, Ruys and De Baere, *International Law: a European Perspective*, n 17 above, 325.

Convention. It is sometimes said that the EU is a 'quasi-member'. While in most of the OECD's work, the Union takes part as a full participant, it is a full member of the Development Assistance Committee (DAC).[67] The Union does not contribute to the budget of the OECD, but makes substantial voluntary contributions to the organisation.[68] While the EU representative is not entitled to vote when legal acts are being adopted by the OECD Council, he or she may be elected as a member of the bureau of subsidiary bodies and participates fully in the preparation of texts, including legal acts, with an unrestricted right to make proposals and suggest changes.[69] Members of the EU Delegation in Paris as well as counsellors and experts from the Commission, EU agencies, and the ECB actively engage in regular dialogue, through the OECD technical committees, in the areas of taxation, agriculture, health, statistics, trade, investment, climate, digitalisation, education, employment, social affairs, innovation, development cooperation, and many other issues.[70]

The EU benefits from the statistical data and economic studies of the OECD, which include surveys of EU policies and of the Eurozone. OECD policy and standard-setting initiatives often find their way into EU legislation, like the 2016 and 2017 anti-tax avoidance directives,[71] which implement into EU law the OECD's recommendations against base erosion and profit shifting (BEPS) to ensure that the profits of multinational enterprises (MNEs) are taxed in the location where the economic activities generating those profits are performed and value is created, as well as the 2022 Council Directive on ensuring a global minimum level of taxation for multinational enterprise groups and large-scale domestic groups in the Union,[72] which enacts in Union law the OECD's 2021 Model Rules ('Tax Challenges Arising from the Digitalisation of the Economy – Global Anti-Base Erosion Model Rules Pillar Two').

[67] Vlastou-Dimopoulou, n 63 above, at 327–28.

[68] EU funds represented a third of all voluntary funding given to the OECD in 2020: https://www.eeas.europa.eu/paris-oecd-unesco/relations-oecd-and-unesco_en?s=64.

[69] https://www.oecd.org/eu/european-union-and-oecd.htm.

[70] https://www.eeas.europa.eu/paris-oecd-unesco/relations-oecd-and-unesco_en?s=64.

[71] Council Directive (EU) 2016/1164 of 12 July 2016 laying down rules against tax avoidance practices that directly affect the functioning of the internal market, OJ L 193/1, 19.7.2016; Council Directive (EU) 2017/952 of 29 May 2017 amending Directive (EU) 2016/1164 as regards hybrid mismatches with third countries, OJ L 144/1, 7.6.2017.

[72] Council Directive (EU) 2022/2523 of 14 December 2022 on ensuring a global minimum level of taxation for multinational enterprise groups and large-scale domestic groups in the Union, OJ L 328/1, 22.12.2022.

VI. Limited Actorness: The EU at UN Economic Fora[73]

As indicated above, the UN is eminently a Westphalian, or state-centric, organisation, where only 'peace-loving States' are admitted to membership.[74] In most (but not all) UN organs, bodies, programmes, funds, and specialised agencies, the EU has been granted an observer status over time. The focus is only on the most important economic fora here.

As one of the objectives of the UN is '[t]o achieve international co-operation in solving international problems of an economic ... character',[75] two of its principal organs are worth examining. The first is the UN General Assembly (UNGA), which the UN Charter tasks to 'initiate studies and make recommendations for the purpose of ... promoting international co-operation in the economic .... fields' (below, section VI.A).[76] The second is the Economic and Social Council (ECOSOC),[77] which is tasked to

> make or initiate studies and reports with respect to international economic ... matters and may make recommendations with respect to any such matters to the General Assembly, to the Members of the United Nations, and to the specialized agencies concerned.[78] [below, section VI.B]

Apart from these principal organs, a number of other UN bodies dealing with economic governance can be mentioned (below, section VI.C).

A. UN General Assembly

In the UNGA, where all UN Member States are represented, the EEC has been an observer since 1974.[79] When the Treaty of Lisbon did away with the external representation role of the rotating Council Presidency – which hitherto had been the Member State speaking on behalf of the EU and its Member States in

[73] This section draws notably from AL Chané and J Wouters, 'The European Union in United Nations Economic Governance Fora' (2017) 8 *European Yearbook of International Economic Law* 541–61.
[74] Article 4(1) United Nations Charter.
[75] Article 1(3) United Nations Charter.
[76] Article 13(1)(b) United Nations Charter.
[77] Article 7(1) United Nations Charter.
[78] Article 62(1) United Nations Charter.
[79] UNGA Res 3208 (XXIX), 'Status of the European Economic Community in the General Assembly', 11 October 1974.

the UNGA – the Union negotiated a more enhanced position for itself in the UNGA. This resulted in the UNGA 'Status Resolution' of 3 May 2011,[80] which gives the Union enhanced participation rights. More concretely, it gives the EU the opportunity to speak early among representatives of other major groups, when speaking on behalf of the 27 EU Member States, and invites it to intervene in the general debate at the opening of the UNGA. In addition, the Union obtained the right to orally present proposals and amendments, and the right to reply once to a speech regarding EU positions. The Union does not have the right to vote, to co-sponsor draft resolutions or decisions, or to put forward candidate members. Moreover, it only applies to the participation of the EU in the sessions and work of the UNGA, its committees and working groups, in international meetings and conferences convened under the auspices of the General Assembly, and in UN conferences.[81] It does therefore not extend the enhanced participation rights of the Union to subsidiary bodies of the UNGA, like the UN Human Rights Council.

Based on the work of its Second Committee, which considers economic and financial matters, the UNGA adopts a number of resolutions every year that bear on global economic governance.[82] While EU Member States have been consistently in favour of consensually agreed UN agendas like the SDGs, they have voted against certain initiatives of the Global South, like resolutions on the 'New International Economic Order'.[83]

[80] UNGA Res 65/276 of 11 May 2011, 'Participation of the European Union in the work of the United Nations', A/RES/65/276. On this resolution see, inter alia, S Blavoukos, D Bourantonis and I Galariotis, 'In Quest of a Single European Union Voice in the United Nations General Assembly' (2017) 52(4) *Cooperation and Conflict* 451; PA Serrano, *Participation of the European Union in the Work of the United Nations: General Assembly Resolution 65/276* (The Hague, CLEER Working Paper 2012/4); J Wouters, T Ramopoulos and J Odermatt, 'The Lisbon Treaty and the Status of the European Union in the International Arena: the May 2011 Upgrade at the UN General Assembly', *IEMed. Mediterranean Yearbook 2011*, 166–70; J Wouters, J Odermatt and T Ramopoulos, 'The Status of the European Union at the United Nations General Assembly', in I Govaere, P Van Elsuweghe, E Lannon and S Adam (eds), *Liber Amicorum Marc Maresceau* (Leiden, Brill Publishers, 2013), 211–23.

[81] Council of the EU, Relations between the European Union and the United Nations, EU-UN / 1 September 2011, https://www.consilium.europa.eu/uedocs/cms_data/docs/pressdata/EN/foraff/124604.pdf.

[82] See for instance, for 2022, the resolutions mentioned in the press report of the UNGA's 77th meeting, G1/12482, 14 December 2022, 'General Assembly Takes Up Second Committee Reports, Adopting 38 Resolutions, 2 Decisions', https://press.un.org/en/2022/ga12482.doc.htm.

[83] See recently, for instance, the voting for UNGA Res 77/174, 'Towards a New International Economic Order' of 14 December 2022, A/RES/77/174. All EU Member States voted against, https://documents-dds-ny.un.org/doc/UNDOC/GEN/N22/751/26/PDF/N2275126.pdf?OpenElement, at 14. The New International Economic Order's (NIEO) dynamics go back, of course, all the way to the 1970s and 1980s. For the then Community's approach to it, see CC Twitchett, 'A New International Economic Order: the European Community's Response' (1978–79) 17 *Journal of Common Market Studies* 156–63.

B. ECOSOC

Unlike the UNGA, ECOSOC is a body with limited membership. Its 54 members are elected by the UN General Assembly for a term of three years, taking into account geographical representation. Relations between ECOSOC and the EEC began on an informal basis and were progressively strengthened in the 1960s and 1970s. After having been admitted to ad hoc participation subject to invitation, the EEC received permanent observer status in 1979.[84] The EU has such status also in the eight functional commissions of ECOSOC.[85] In terms of the regional commissions of ECOSOC, the EU's participation as an observer in the work of the UN Economic Commission for Europe (UNECE) is especially important to note.[86] The EU has long implemented in its legislation technical regulations adopted by UNECE, and is actively involved in the discussions thereon.[87] Apart from that, the Union (thanks to an REIO clause[88]) and its Member States are parties to a number of multilateral treaties adopted by UNECE, such as the Convention on Long-Range Transboundary Air Pollution and the Convention on Access to Information, Public Participation in Decision-making and Access to Justice in Environmental Matters (Aarhus Convention).[89]

[84] See F Hoffmeister and PJ Kuijper, 'The Status of the European Union at the United Nations: Institutional Ambiguities and Political Realities', in J Wouters, F Hoffmeister and T Ruys (eds), *The United Nations and the European Union: An Ever Closer Partnership* (The Hague, TMC Asser Press, 2006), 9, at 19–20.

[85] These are the Commission on Narcotic Drugs, Commission on Population and Development, Commission on Science and Technology for Development, Commission on the Status of Women, Statistical Commission, Commission for Social Development, Commission on Crime Prevention and Criminal Justice, and the Forum on Forests.

[86] For the relevant UNECE Decision of 15 April 1975 admitting the EEC as a regular observer, see Commission of the European Communities, *The European Community, International Organizations and Multilateral Agreements*, 3rd edn (Luxembourg, Official for Official Publications of the European Communities, 1983), 43.

[87] See eg Council Decision (EU) 2022/2200 of 8 November 2022 on the position to be taken on behalf of the European Union in the World Forum for Harmonization of Vehicle Regulations of the United Nations Economic Commission for Europe on proposals for modifications to UN regulations Nos 0, 13, 24, 34, 43, 48, 67, 83, 90, 118, 125, 127, 129, 149, 151, 158, 159, 161, 162 and 163, on a proposal for a new UN Regulation on Vulnerable Road Users in Front and Side Close Proximity, on a proposal for a new UN Regulation on the Approval of Motor Vehicles with Regard to their Direct Vision, and on a proposal for an amendment to M.R.1, OJ L 292/54, 11.11.2022.

[88] The REIO clause stands for 'regional economic integration organisation'. It is typically a clause in multilateral treaties that allows the EU to be a party. See J Odermatt, *International Law and the European Union* (Cambridge, Cambridge University Press, 2021), 70–75.

[89] See Council Decision 2005/370/EC of 17 February 2005 on the conclusion, on behalf of the European Community, of the Convention on access to information, public participation in decision-making and access to justice in environmental matters, OJ L 124/1, 17.5.2005. On the 'Aarhus Regulation' and its 2021 amendments, see https://environment.ec.europa.eu/law-and-governance/aarhus_en.

A special case is the EU's participation in the UN High-level Political Forum on Sustainable Development (HLPF). This forum replaces the former Commission on Sustainable Development (CSD), and the EU has retained in it the 'full participant' status that it enjoyed in the CSD.[90] In July 2023 the Union carried out its first voluntary review on the implementation of the 2030 Agenda for Sustainable Development.[91]

C. Other UN Bodies

In other UN bodies that deal with economic governance, the EU typically only participates as an observer. A case in point is the UN Conference on Trade and Development (UNCTAD). UNCTAD, established in 1964 by the UNGA as a subsidiary body, was for a long time a rather important actor in global economic governance. The establishment and dynamics of the WTO since 1995 have somewhat cast a shadow on it.[92] UNCTAD has been tasked with promoting international trade and economic development, policy formulation, and initiating the negotiation of international legal instruments in the field of trade.[93] The EU participates as an observer in the sessions of the Conference and the UNCTAD Trade and Development Board, as well as the subsidiary organs and working groups.[94] In the latter, the EU is de facto able to participate as a full participant.[95]

UNCTAD was at the origin of the Generalized System of Preferences (GSP) in the 1960s, which has been implemented by the EU (then EEC) since 1971.

[90] See UNGA Res 67/290, 'Format and organizational aspects of the High-Level Political Forum on Sustainable Development', 9 July 2013, A/RES/67/290, para 10.

[91] See European Union, *EU Voluntary Review on the Implementation of the 2030 Agenda for Sustainable Development* (2023), https://hlpf.un.org/sites/default/files/vnrs/2023/VNR%202023%20EU%20Report.pdf.

[92] Through a 2003 Memorandum of Understanding, the WTO and UNCTAD have a strategic partnership for the purpose of implementing the DDA, cooperating to ensure that trade serves development goals, and assisting the beneficial integration of the developing and least developed countries into the global economy and the multilateral trading system, https://www.wto.org/english/thewto_e/coher_e/wto_unctad_e.htm.

[93] UNGA Res 1995(XIX), 'Establishment of the United Nations Conference on Trade and Development as an organ of the General Assembly', UN Doc A/RES/1995(XIX), 30 December 1964, Article II(3).

[94] See UNCTAD, List of intergovernmental organizations participating in the activities of UNCTAD, TD/B/IGO/LIST/12, 14 August 2023, https://unctad.org/system/files/official-document/tdigolistd12_en.pdf.

[95] See M Emerson, R Balfour, T Corthaut, J Wouters, PM Kazynski and T Renard, *Upgrading the EU's Role as Global Actor*, CEPS Commentary, 25 January 2011, www.ceps.eu, at 78; J Scheffler, *Die Europäische Union als rechtlich-institutioneller Akteur im System der Vereinten Nationen* (Heidelberg, Springer, 2011), at 446 and 619.

Furthermore, the EU is also a contracting party to a number of commodities agreements that came about under the auspices of UNCTAD. Initially, these negotiations were only open to states, while the EEC was merely granted consultative status, despite its far-reaching competences in these areas. Since the 1960s this has gradually changed[96] and today the Union is a member of several commodities organisations, sometimes to the exclusion of EU Member States.[97]

There are two UN bodies in the economic field where the EU has become a 'Member Organization': the FAO (since 1991) and the Codex Alimentarius Commission (since 2003), both thanks to an REIO clause.[98]

VII. (Almost) No Actorness: The EU at the Bretton Woods Institutions

The EU does not have any proper status at the Bretton Woods institutions (BWI), ie the International Monetary Fund (IMF or Fund) and the World Bank. As each of these institutions touches on different competences of the EU, they are dealt with successively.

A. The EU at the IMF[99]

In spite of the fact that 'monetary policy for the Member States whose currency is the euro' is an exclusive competence of the Union (Article 3(1)(c) TFEU), the

[96] See Scheffler, n 95 above, at 240–50.

[97] This is the case, eg, for the International Coffee Organization (only the EU: see https://icocoffee.org/what-we-do/about-us/), the International Cocoa Organization (https://www.icco.org/who-we-are/membership/#toggle-id-2) and the International Olive Council (https://www.internationaloliveoil.org/about-ioc/list-of-ioc-members/, which site mentions: 'Members of the European Union are part of the IOC without the need of an application by the country itself.').

[98] See Chapter 5 by Frank Hoffmeister in this volume. See also B van der Meulen and B Wernaart, 'Food and Agriculture Organization (FAO) and Codex Alimentarius Commission. The Impact of the Codex Alimentarius and the Right to Food on EU Food Law', in Wessel and Odermatt (eds), *Research Handbook on the European Union and International Organizations*, n 43 above, 82–100.

[99] This section is based on J Wouters and S Van Kerckhoven, 'The International Monetary Fund', in K E Joergensen and KV Laatikainen (eds), *Routledge Handbook on the European Union and International Institutions: Performance, Policy, Power* (London, Routledge, 2012), 221–33; J Wouters, S Van Kerckhoven and T Ramopoulos, 'The EU and the Euro Area in International Economic Governance: The Case of the IMF', in F Amtenbrink and D Kochenov (eds), *The European Union's Shaping of the International Legal Order* (Cambridge University Press, 2013), 306–27; J Wouters and T Ramopoulos, 'Time to Reconsider Status: the IMF, the EU, the Euro Area and Its Sovereign Debt Crisis', in M Guderzo and A Bosco (eds), *A Monetary Hope for Europe. The Euro and the Struggle for the Creation of a New Global Currency* (Florence, Firenze University Press, 2016), 21–37.

EU does not enjoy any formal status in the IMF. Representation in the Fund takes place via two EU institutions, the ECB and the Commission, as well as by its Member States. The ECB has observer status at the IMF Executive Board and can express its views about European monetary matters during the Board's meetings. The Commission has observer status at the 'Joint Ministerial Committee of the Boards of Governors of the World Bank and the International Monetary Fund on the Transfer of Real Resources to Developing Countries', usually referred to as the Development Committee. Given their limited status in the Fund, the two institutions have a rather circumscribed role in the work of the IMF. The responsibility of representing the EU in all other functions of the IMF, therefore, falls on the shoulders of EU Member States, which are all full member countries of the Fund. However, EU Member States are dispersed within the Fund in different constituencies, sometimes comprising a combination of euro area Member States, non-euro area Member States and third countries. This leads to a diminished visibility and, arguably, influence in the Fund, despite the fact that EU Member States jointly hold a quota of more than 30 per cent. In order to tackle this problem, two coordination mechanisms exist in Washington and Brussels, respectively. The group of EU representatives to the IMF (EURIMF) convenes in Washington, whereas the Sub-Committee on IMF (SCIMF) operates under the ECOFIN Council in Brussels. However, coordination through these two mechanisms is considered insufficient and often fails to produce common EU positions and, ultimately, statements in the Fund. The international financial crisis and the euro area's sovereign debt crisis have worsened this problem, since they have often laid bare the strongly diverging national interests among EU Member States. At the same time, Member States have been reluctant to see a further strengthening of EU representation in the Fund through unification, since this would come at the expense of having their own voices heard. The 'Five Presidents Report' of 2015 called for unification, stating that

> in the international financial institutions, the EU and the euro area are still not represented as one. This fragmented voice means the EU is punching below its political and economic weight as each euro-area Member State speaks individually. This is particularly true in the case of the IMF despite the efforts made to coordinate European positions.[100]

Later that year, the Commission proposed a Council decision 'laying down measures in view of progressively establishing unified representation of the euro area in

[100] JC Juncker et al., *Completing Europe's Economic and Monetary Union* (European Commission, 2015), https://commission.europa.eu/system/files/2016-03/5-presidents-report_en.pdf, at 17.

the International Monetary Fund' based on Article 138 TFEU;[101] however, it did not attract sufficient support in the Council.

B. The EU at the World Bank[102]

Despite being a major donor of Official Development Assistance in general and ranking third among the contributors to World Bank trust funds, the EU has no formal status in the World Bank. As indicated above, the Commission has observer status in the Development Committee. Like the IMF, the World Bank only recognises states as members. The Member States' ministers of finance or development sit on the World Bank's Board of Governors, and they appoint the World Bank's 25 Executive Directors, who, together with the President of the World Bank Group, make up the Board of Directors. Two EU Member States (France and Germany) are among the World Bank's largest members and may consequently appoint their own Executive Director. The other Member States are spread across different constituencies.

The relevant competence of the EU here is development policy, which is a 'parallel' shared competence of the EU and its Member States (Article 4(4) TFEU). While the exercise of the competence by the Union does not prevent the Member States from exercising their competences, the principle of sincere cooperation requires them to refrain from any action that may run counter to obtaining the EU's objectives.

EU coordination at the World Bank is not very developed. There is no permanent structure of coordination in Brussels. Coordination takes place in Washington between the EU Executive Directors and EU Member State representatives. In 2003, under the Italian Presidency, the Italian Executive Director established an informal consultation process of the EU Executive Directors in Washington to coordinate European positions on the Board of Directors, exchange information, and agree on joint positions. The EU Delegation in Washington participates in these meetings as an observer. Coordination is hampered in particular by the lack

[101] European Commission, Proposal for a Council decision laying down measures in view of progressively establishing unified representation of the euro area in the International Monetary Fund, 21 October 2015, COM(2015) 603 final. In 1998 the Commission had already made an unsuccessful first proposal: European Commission, Proposal for a Council Decision on the representation and position taking of the Community at international level in the context of Economic and Monetary Union, 9 November 1998, COM(98) 637 final.

[102] This section is based on AL Chané and J Wouters, above n 73, at 557–59.

of a permanent forum in Brussels, the disparate representation of EU Member States in the World Bank (with ministers covering a diverse range of portfolios), the quick reaction times necessitated by World Bank projects, and the governance structure of the World Bank, which disperses EU Member States across different constituencies.

Nevertheless, the EU has developed a strong partnership with the World Bank, as indicated by the EU's financial contributions to the World Bank's funds. In 2001, both organisations concluded a 'Trust Funds and Co-Financing Framework Agreement', subsequently renewed and revised in 2009, 2014, 2016, and 2020,[103] which governs the use of EU budget money for World Bank development projects. The European Commission and the World Bank exchange information and coordinate on development cooperation, conduct policy dialogues on trade and agriculture, and cooperate in the area of global public goods, such as environmental protection, control of communicable diseases such as HIV/AIDS, and preventing or mitigating crises in the international financial system. However, the gap between the EU's financial contributions[104] and its representation in the World Bank persists. Proposals to combine the various EU Member State Executive Directorates into a single EU seat on the Board of Directors have so far received little support from EU Member States.

VIII. Concluding Remarks

This chapter has taken a closer look into the EU's role and position in some of the main institutions of global economic governance. It started with informal clubs, notably the G7 and G20, and continued with formal economic organisations. The latter were studied in a deliberate order, namely from the WTO, where the EU enjoys full membership rights, onwards to the OECD, where the Union acts as a full participant, to UN economic fora, where the EU typically has to work from a more modest observer status, to the Bretton Woods institutions, where the Union presently has no status at all.

[103] See 'European Commission and World Bank Group renew agreement to strengthen development cooperation', Commission press release, 8 July 2020, https://ec.europa.eu/commission/presscorner/detail/en/ip_20_1300.

[104] For the period 2015–19, the EU contributed €2.07 billion to World Bank Group trust funds – including the World Bank, the International Finance Corporation (IFC), and the Multilateral Investment Guarantee Agency (MIGA): ibid.

From this overview, it is clear that the Union actively participates in most global economic governance institutions, but that this happens in a rather uneven manner, depending on the EU's (offensive / defensive) interests, its competences, its external representation, and the institutional position of itself and its Member States, as well as the external context, ie the rules governing membership and participation in the institutions, as well as the attitudes of participating third countries.

What is most striking is that the EU currently does not seem to have any comprehensive strategy to upgrade its position in those institutions where improvement would make sense. In 2012 then European Commission President, José Manuel Barroso, and Vice-President, Catherine Ashton, proposed a rather modest 'Strategy for the progressive improvement of the EU status in international organisations and other fora in line with the objectives of the Treaty of Lisbon'.[105] However, in practice, EU efforts for status upgrades have faced serious political resistance, both internally and externally. One can only hope that one day, EU institutions and Member States get their act together and develop a much more holistic vision and action plan on how to optimise their scattered presence and role in the various institutions of global economic governance.

[105] Communication to the Commission from the President in Agreement with Vice-President Ashton, *Strategy for the progressive improvement of the EU status in international organisations and other fora in line with the objectives of the Treaty of Lisbon*, Brussels, 20 December 2012, C(2012) 9420 final.

7

The European Union and the UN Human Rights Council: Reconciling European and International Efforts to Protect Human Rights Worldwide

CHRISTINA KOKKINAKIS AND
ANDREJ AUERSPERGER MATIĆ

I. Introduction

Once a year, shortly before the main session of the United Nations Human Rights Council in Geneva starts, the 27 ministers of foreign affairs of the European Union (EU) adopt Council Conclusions on EU priorities in United Nations (UN) human rights fora. With these Council Conclusions, the EU reaffirms its commitment to respect, protect and fulfil all human rights and, most importantly, to defend their universality. The year 2023 was particularly relevant in this regard, given that it marked the seventy-fifth anniversary of the Universal Declaration of Human Rights (UDHR), which coincided with the thirtieth anniversary of the Vienna Declaration and Programme of Action (VDPA) adopted in June 1993. Together with the international community, the EU reaffirmed that human rights are universal, indivisible, interdependent and interrelated, be they economic, social, cultural, civil or political.

For the EU, the United Nations Human Rights Council (HRC) has been, together with the Third Committee, the most appropriate international forum to raise its voice against human rights violations and abuses, wherever they occur. Indeed, the EU and the UN can be said to have a symbiotic relationship in this area: the EU 'exports' its human rights values through the UN, while the latter needs EU support to be able to reinforce human rights objectives worldwide.[1]

However, the EU is not a state and therefore not a member of the Human Rights Council. How does it, then, participate in and interact with the UN Human Rights Council when it comes to specific issues? This question requires an examination of two distinct issues. First, what precisely is the EU's status in the HRC? Second, how do internal legal developments happening in the EU, itself an organisation of closely associated states, affect the debate in the HRC? This contribution seeks to shed some light on these two issues, by situating the HRC in the broader discussion on the relationship between the EU and the UN, and by providing a case study that neatly demonstrates the intricacies of the interaction between the two international systems when they elaborate internationally binding legal rules.

II. Background

The position of the EU in the UN system and its various organs and bodies differs significantly depending on the nature of the representation that is required and even possible in the given institutional framework. In the UN Security Council (UNSC), for instance, the position of the EU has not been formalised at all. The EU is, as are other non-members of the UNSC, regularly invited to speak in the UNSC under Rule 39 of its Rules of Procedure, whenever the nature of the meeting allows it. The EU High Representative for Foreign Affairs and Security Policy addresses the UNSC regularly, but there is no particular legal rule that would grant either him or her, or indeed the EU, the right to appear and make representations to Council members.

A different and very interesting path was taken in the United Nations General Assembly (UNGA), which granted, in October 1974, observer status to the then European Economic Community, which thus became the first non-state entity to obtain such status.[2] This marked the beginning of the representation of what is now the EU in the assembly of the most important international political forum. Subsequently, at the beginning of the new millennium and concurrently with the

[1] M Schmidt, 'The Position of the European Union in the United Nations: a United Nations Perspective', in C Kaddous (ed), *The European Union in International Organisations and Global Governance: Recent Developments* (Oxford, Hart Publishing, 2015) 33, at 36.

[2] UNGA Res A/RES/3208(XXIX).

development of its own institutional architecture, the EU sought to strengthen its status in multilateral fora (and UN bodies in particular) from a mere observer to an active participant. After the entry into force of the Lisbon Treaty in 2009, one of its main objectives was to secure active participation rights in the UNGA. After difficult negotiations, on 3 May 2011, UNGA adopted Resolution 65/276 entitled 'Participation of the European Union in the work of the United Nations'.[3] This resolution reaffirmed the principle that membership is 'limited to States that are Members of the United Nations' but recognised the specificity of the EU among the observers. In view of that specificity, the resolution defined special modalities

> for the participation of the representatives of the European Union, in its capacity as observer, in the sessions and work of the General Assembly and its committees and working groups, in international meetings and conferences convened under the auspices of the Assembly and in United Nations conferences.

Resolution 65/276 upgraded the role of the EU and its status in UNGA in line with the Treaty of Lisbon, granting the EU a set of specific procedural rights: (i) to be 'inscribed on the list of speakers among representatives of major groups', so as to make interventions; (ii) to be 'invited to participate in the UNGA general debates in accordance with the order of precedence as established in the practice for participating observers and the level of participation'; (iii) to exercise the right of reply where the EU has expressed own positions as decided by the presiding officer, such right of reply being restricted to one intervention per item; (iv) to present proposals and amendments orally as agreed by the EU Member States and (v) to have its communications circulated directly as documents of the Assembly, meeting or conference concerned. The resolution did not, on the other hand, grant the EU the right to vote, to co-sponsor draft resolutions or decisions, to put forward candidate states, or to raise points of order.

A note by the UN Secretary-General provided further details on the modalities of participation and the rationale.[4] It recalled the reasoning of the resolution to the effect that

> States members of the European Union had entrusted the external representation of the European Union, previously performed by the representatives of the member State holding the rotating Presidency of the Council of the European Union, to the

[3] UNGA Res 65/276, Participation of the European Union in the Work of the United Nations, 3 May 2011.

[4] Note by the Secretary-General, 'Participation of the European Union in the work of the United Nations', A/65/856.

> following institutional representatives: the President of the European Council, the High Representative of the Union for Foreign Affairs and Security Policy, the European Commission and the European Union delegations, which had assumed the role of acting on behalf of the European Union in the exercise of the competences conferred by its member States.

The note also clarified that Resolution 65/276 does not apply to the other principal organs of the United Nations or to their subsidiary organs and is 'without prejudice to any privileges of participation which have been or may be conferred upon the European Union by them'. This view therefore allows for special arrangements to be made in various subsidiary organs and bodies, taking into account their specificities and structure.

III. The Position of the EU at the HRC

The EU was among the most important actors behind the creation of the HRC in 2006 and continues to be a strong supporter of its work and its various reporting mechanisms.[5] In terms of decisional procedures, the practice of the HRC, one of the most important UN bodies, is distinct from both the UNSC and the UNGA. Its character and composition almost by definition limit the possibilities for the EU to act as a full participant. Its Rules of Procedure, adopted in a 2007 resolution,[6] provide that the HRC 'shall apply the rules of procedure established for the Main Committees of the General Assembly, as applicable, unless subsequently otherwise decided by the Assembly or the Council'.[7] Regarding the participation of observers, it is stated that such participation shall be based on arrangements, including Economic and Social Council resolution 1996/31 of 25 July 1996 (concerning non-governmental organisations), and 'practices observed by the Commission on Human Rights, while ensuring the most effective contribution of these entities'.[8] It should be noted however, that the unsatisfactory position of the EU in the HRC was recognised early on. Most importantly, the 'Barroso–Ashton' Strategy, a 2012 Communication to the Commission,[9] recommended enhancing the formal status

[5] Schmidt, n 1 above, at 37.
[6] Resolution 5/1, 'Institution-building of the Human Rights Council', A/HRC/5/L.2.
[7] HRC Rules of Procedure, Rule 1.
[8] HRC Rules of Procedure, Rule 7.
[9] Communication to the Commission from the President in Agreement with Vice-President Ashton – Strategy for the progressive improvement of the EU status in international organisations and other for a in line with the objectives of the Treaty of Lisbon, C(2012) 9420 final, 20 December 2012.

of the EU in the HRC by applying the principles of UNGA Resolution 65/276 to subsidiary organs, focusing on the HRC in particular. This objective has thus far not been achieved, and it poses a number of questions.[10] In particular, existing arrangements make it unnecessarily difficult for the EU to formulate positions on behalf of its Member States, which is rather problematic given the importance of the EU as a preeminent defender of human rights and the particular significance attached to international human rights protection in Europe.

At present, voting rights in the HRC are reserved for members, in accordance with Rule 124 of the UNGA Rules of Procedure. The EU also cannot formally table any draft resolution or written amendment to a draft resolution in the name of the EU, and cannot put proposals and amendments to a vote in the HRC. Resolutions proposed by the EU are therefore tabled and introduced (before action) by the rotating EU Presidency on behalf of the EU. As for the right of the EU to raise points of order, it is perhaps interesting that a provision to that effect was included in the initial draft resolution on the status of the EU in UNGA. It was presented as a proposal but was ultimately left out of the final text of Resolution 65/276 as a consequence of a concession by the EU at the end of a difficult negotiation process.

Secondly, while Rule 113 of the UNGA Rules of Procedure grants any 'representative' the right to raise a point of order, this rule applies only to representatives of one of the UN's members and not to those of an observer. While at UNGA general debates the EU Delegation to the United Nations in New York speaks on behalf of the EU, at the HRC general debates it is the rotating EU Presidency or any other EU Member State (that is also an elected member of the HRC) that delivers statements. In the absence of clear rules, the question of whether the EU is entitled to speak may depend on the relevance of the relevant resolution for the EU, as well as existing HRC practices. By contrast, during HRC interactive dialogues and debates, high-level fora, intersessional events and open-ended working groups, it is the EU Delegation that speaks on behalf of the EU as a whole. At the Universal Periodic Review, which is a peer review mechanism between states, the EU participates as an observer, including in the interactive dialogues. The EU expresses its views regarding an outcome of a review before the plenary takes action on it.

Ultimately, the interpretation of the rather unclear set of rules and their application to the HRC is a matter for HRC members to decide. That said, it is clear that

[10] J Wouters, A Chané, J Odermatt and T Ramopoulos, 'Improving the European Union's Status in the United Nations and the UN System: An Objective Without a Strategy?', in C Kaddous (ed), *The European Union in International Organisations and Global Governance: Recent Developments* (Oxford, Hart Publishing 2015), 45, at 55–59.

any development of participation rights that might be agreed by the HRC would be without prejudice to the EU's current practice of exercising its rights at HRC sessions in agreement with its Member States. Indeed, in practical terms, strong coordination has been set up over the years among the EU Member States to align the positions expressed in the HRC, especially by way of internal meetings before regular and special sessions of the HRC.[11] As the latter is based in Geneva, coordinating the EU position falls under the responsibility of the EU Delegation to the United Nations in Geneva, which has assumed tasks that were exercised previously by the EU Presidency. The work of the HRC is part of the regular exchanges with the EU Heads of Mission in Geneva, and it is precisely in the area of human rights that EU coordination is very intensive.[12] The work of the EU Delegation in Geneva is conducted consistently with policy orientation formulated at the European External Action Service headquarters in Brussels, taking due account of the policy positions expressed by the EU Member States in Council.

IV. Formulating Legal Instruments

How does the EU manage complex engagement in the HRC given that it is itself a highly complex legal structure, with competences in areas concerning human rights split between the Member State and EU levels? When formulating positions on various statements, reports or mechanisms, the exercise of the EU competence does not pose problems other than political coordination, as the EU Member States may have different approaches to the issues discussed, especially if there is a strong foreign policy dimension. On the other hand, a particularly interesting situation arises where the HRC and the EU seek to elaborate legally binding rules on the same subject-matter.

A. The Human Rights Due Diligence Instrument

The juxtaposition of two distinct legal regimes, their processes and styles is well illustrated in the case of the EU's involvement in negotiations on a proposed UN treaty establishing human rights due diligence obligations for businesses.

[11] Schmidt, n 1 above, at 37.

[12] M Zappia, 'The United Nations: a European Union Perspective', in C Kaddous (ed), *The European Union in International Organisations and Global Governance: Recent Developments* (Oxford, Hart Publishing, 2015) 25, at 29.

The proposed instrument is intended to establish (i) mandatory international human rights due diligence rules for business, as well as (ii) rules on legal liability, jurisdiction, applicable law and procedure, aimed at ensuring effective access to justice for victims of human rights violations, applicable mainly to transnational corporations. Thus far, EU engagement at the negotiations has been limited due to the lack of negotiation directives, as well as due to serious reservations about the aims and the overall process of negotiations. At the time of writing, these protracted negotiations have already taken ten years since their inception and there is no clear indication that they might be completed any time soon. From the political point of view, a key factor behind the slow progress is the radical divergence of views among industrialised and developing countries regarding the feasibility, appropriate scope and content of the proposed instrument.

This controversy, and the debate on whether and to what extent to apply human rights obligations to businesses internationally, did not come out of nowhere. Attempts to ensure accountability of large foreign businesses for human rights violations occurring in third (mainly developing) countries in fact have a history spanning several decades, becoming more prominent with widespread publicity about the violence and pollution in the Amazon and the Niger delta. In recent years, public pressure to ensure accountability resulted in the formation of several non-binding documents, the most significant of which was the UN Guiding Principles on Business and Human Rights (UNGP),[13] adopted unanimously by the HRC in 2011. The UNGP 'Protect, Respect and Remedy' framework has since then been the main frame of reference for voluntary mechanisms seeking to ensure compliance of international business with human rights. In essence, it envisages due diligence mechanisms (ie self-policing to prevent the occurrence of adverse human rights impact, and mitigation measures if they occur), applicable to all businesses, and recommends ways to improve access to remedies for affected victims (through a combination of judicial and non-judicial grievance mechanisms).

In 2014, Ecuador and South Africa proposed to take a further step and tabled a resolution at the HRC with a view to beginning a process towards an international legally binding treaty on business and human rights. Agreeing to take up the idea, the HRC adopted Resolution 26/9 establishing

> an open-ended intergovernmental working group [OEIWG] on transnational corporations and other business enterprises with respect to human rights, with a mandate

[13] United Nations Office of the High Commissioner for Human Rights, 'Guiding Principles on Business and Human Rights: Implementing the United Nations "Protect, Respect and Remedy" Framework', HR/PUB/11/04 (New York and Geneva, 2011).

> to elaborate an international legally binding instrument to regulate, in international human rights law, the activities of transnational corporations and other business enterprises.

The EU Member States and other industrialised countries did not participate in the vote on this resolution. It transpired from the vote and the associated debate that from the viewpoint of developing countries (and the vast majority of international NGOs) an international instrument governing the accountability of businesses should be as broad as possible and allow victims a wide array of legal remedies in countries other than the one where the human rights violations took place. Moreover, they maintained that the focus of the international obligations should be transnational corporations. Industrialised countries and business associations, on the other hand, tended to see such proposals as excessively broad and contrary to the principles of legal certainty, as well as ideologically biased against foreign investors.

B. The Process of Deliberation

In 2018, the OEIWG presented the 'Zero Draft' of a 'Legally Binding Instrument to Regulate, in International Human Rights law, the Activities of Transnational Corporations and Other Business Enterprises' (LBI), prepared by the Chair Rapporteur, which also included a draft optional protocol.[14] This was followed by several other drafts, the most important one being the 'Third Revised Draft' circulated in 2021.[15] The Third Revised Draft and the concrete textual proposals submitted by states during the seventh session was the basis for discussions at the eighth session of the OEIWG in October 2022. The eighth OEIWG session was more constructive compared to previous years, but it remained obvious that there is a clear divergence of views among participants. On the one hand, the so-called 'Global South', supported by most civil society organisations, remained the dominant voice, stressing that the scope of the future LBI should be restricted to transnational corporations, and demanding stronger and detailed provisions to hold businesses accountable. On the other hand, industrialised countries,

[14] Zero Draft of a Legally Binding Instrument to Regulate, in International Human Rights law, the Activities of Transnational Corporations and Other Business Enterprises, 16 July 2018.

[15] Third Revised Draft of the Legally Binding Instrument to Regulate, in International Human Rights law, the Activities of Transnational Corporations and Other Business Enterprises, 17 August 2021, https://www.ohchr.org/sites/default/files/LBI3rdDRAFT.pdf (Third Revised Draft).

including the EU Member States and organisations representing businesses, reiterated their call to move towards a less prescriptive and more realistic approach, covering all companies (including domestic and state-owned companies) in order to ensure a level playing-field. They also emphasised that wider participation by states was needed in order for the LBI to be effective. In order to bridge the gap between positions, and promote engagement by industrialised countries in particular, a month before the meeting the Chair-Rapporteur shared suggested proposals for amendments to several articles. His attempt to find a compromise was rejected and the Chair-Rapporteur therefore reverted to the 'Group of Friends' (which included EU Member States France and Portugal), giving responsibility to this limited group for making substantive progress.

In July 2023, the Chair-Rapporteur circulated a Fourth ('Updated') Revised Draft,[16] which took into account some of the main objections raised by the industrialised countries, in particular regarding the scope of the instrument, which was enlarged to cover all companies, both domestic and transnational. The Fourth Revised Draft was accepted, after an initial controversy, as the basis for discussions at the ninth session of the OEIWG in October 2023. The discussions at the ninth session covered only three articles (out of 24) of the text, as the polarised positions persisted amongst the participants over the key aspects of the LBI. As before, several states from the 'Global South' reiterated their call to have the scope limited to transnational corporations and other business enterprises of a transnational character, pointing to the mandate of Resolution 26/9, and insisted on reverting to the Third Revised Draft as the basis of negotiations. On the other hand, certain Latin American countries, as well as European and North American states, supported a wider scope for the LBI as reflected in the Fourth Revised Draft. While the EU coordination process in Geneva continued to be rather intensive, the EU delivered a constructive opening statement signalling, for the first time, that the adoption of the Corporate Sustainability Due Diligence Directive (CSDDD) could provide a basis for the EU's future engagement in the negotiations. A number of EU Member States (Portugal, France and Germany) also delivered opening statements, and two members of the European Parliament participated on the first day of the session.

The relationship between the EU and the HRC in this context was conditioned by the internal dynamics of the EU, the fact that the EU has a specific role in relation

[16] Updated draft legally binding instrument (clean version) to regulate, in international human rights law, the activities of transnational corporations and other business enterprises, July 2023, https://www.ohchr.org/sites/default/files/documents/hrbodies/hrcouncil/igwg-transcorp/session9/igwg-9th-updated-draft-lbi-clean.pdf (Fourth Revised Draft).

to the UN and the HRC in particular, and the divisive politics of the LBI deliberations themselves. Crucially, the EU did not formally take part in LBI negotiations at the eighth and the ninth OEIWG sessions, as it did not have a proper mandate to do so. This was because the EU Council of Ministers and its preparatory bodies had not authorised the Commission and the EEAS to take part and had not issued the necessary negotiating directives. The lack of a mandate to formally engage is perhaps understandable given that internal legislation on human rights due diligence had not yet been adopted by the EU itself for its Member States. However, the Council did indicate openness to discuss the issue, as it adopted, on 24 January 2022, its Conclusions on EU Priorities in UN Human Rights Fora in 2022. In this document, it declared that the EU

> will engage actively in the UN discussions, on a legally binding instrument on business and human rights, including through a possible involvement in the suggested Group of Friends of the Chair and by working with partners to explore ideas for a consensus-based instrument that can effectively enhance the protection of victims and create a global level playing field.[17]

At the eighth and ninth OEIWG sessions, the EU (represented by the EEAS) accordingly made substantial comments on the Third Revised Draft and the first three articles of the Fourth Revised Draft respectively, without being formally committed to negotiations. It was able to do because on 23 February 2022 the Commission proposed a related internal legislative act, namely the draft Corporate Sustainability Due Diligence Directive,[18] which was ultimately adopted on 13 June 2024.[19] This approach was reflected in the discussions held on 28 September 2022 and on 4 October 2023 in the Council Working Party on Human Rights (COHOM), which indicated support for delivering a constructive EU statement. At the 4 October 2023 meeting, COHOM also asked the EU to indicate that the EU might join the negotiations based on its own due diligence legislation.

[17] Council Conclusions on EU Priorities in UN Human Rights Fora in 2022, approved by the Council at its 3844th meeting held on 24 January 2022, point 26, https://www.consilium.europa.eu/media/53956/st05277-en22.pdf.

[18] Proposal for a Directive of the European Parliament and of the Council on Corporate Sustainability Due Diligence and amending Directive (EU) 2019/1937, COM(2022) 71.

[19] Directive (EU) 2024/1760 of the European Parliament and of the Council of 13 June 2024 on corporate sustainability due diligence and amending Directive (EU) 2019/1937 and Regulation (EU) 2023/2859 (OJ L 2024/1760 5.7.2024) (EU Directive).

C. Policy Interaction

The question of whether to formally join the deliberations of the LBI in the HRC working group posed a serious political dilemma for the EU and many of its Member States. In policy terms, political pressures to ensure legal accountability of large companies worldwide had to be balanced against genuine concerns about the possible consequences and the perceived lack of symmetry in international relations. In the light of the adoption of the EU Directive, which enhances the previously existing sectoral legal framework for supply chain due diligence,[20] the EU can be said to have, by definition, an interest in promoting its regulatory model and ensuring a level playing-field for businesses, both inside and outside the EU. Due diligence obligations may impose important regulatory burdens and represent serious legal risks for companies, so it would seem reasonable that efforts in this direction are coordinated among the large economies of the world. Indeed, since 2022 the majority of EU Member States, under pressure from their civil society and parliaments, appeared to favour the EU joining the negotiations. The momentum further increased in 2023, after both the Council and the European Parliament defined their positions on the EU Directive, indicating that its adoption was very likely. The proponents of the LBI consequently argued that it is better for the EU to be part of the negotiations and steer them in the direction of EU rules before it is too late, especially if other participants agree on provisions that differ from the solutions adopted within the EU. Moreover, the EU also has an interest in actively steering the discussions on access to justice to ensure that any provisions would be in line with its *acquis* and consistent with the positions expressed in the context of the EU Directive. In addition, joining the negotiations could allow the EU to express the concerns of EU business representatives, who expressed opposition to broad jurisdictional and liability regimes operating across third countries that might sign up to the LBI, especially in the case of weak legal systems that do not respect basic rule-of-law standards.

On the other hand, several arguments also urged against joining the negotiations. Most importantly, it was not clear at all whether the LBI deliberations would ultimately reach any conclusion, especially given the lengthy discussion process thus far. The text of the draft LBI on the table would in any event need to be revised substantially to align it with EU proposals. In theory, the EU could condition its

[20] F Hoffmeister, 'The European Regulatory Approach on Supply Chain Responsibility' (2022) 25 *Zeitschrift für Europarechtliche Studien* 221–50.

participation by asking for a more fundamental 'reset' in the negotiating process, to consider an alternative, more realistic approach, which would be able to garner support from industrialised countries and representatives of the business sector. Nevertheless, in this scenario, opposition from the countries of the 'Global South' as well as representatives of international NGOs was likely, as the latter opposed any dilution of the text under discussion. As the ninth OEIWG session demonstrated, many proponents of the LBI insist that it should only cover transnational companies and exempt domestic ones, a position that is difficult to accept for countries with advanced economies, as such a differentiation is hard to justify from both the level playing-field and the human rights perspective.

With the discussions going ahead in both Brussels and Geneva, arguments for more formal engagement became stronger. Indeed, being active in the deliberations of the HRC or its working groups that may have strong legal implications is, in one sense, a necessity for the EU, as its own legislative mechanisms may create obligations comparable or overlapping with those that may be formulated in the UN framework. But where political trends lead to legal (as opposed to political) initiatives at both the EU and UN level, a key institutional question emerges: who is ultimately competent to engage with the UN?

This is the perennial issue of EU competence – specifically, external EU competence – in relation to its constituent Member States, which is something of a permanent challenge for the EU as it evolves and develops its internal legislation. In particular, due to the broad character of human rights and the many policy fields that might be impacted, overlapping commitments imposed by EU rules and any UN instrument could lead to contradictions or at the very least to genuine confusion. It should be recalled in this connection that while EU competence in external relations can be based on express conferral in the Treaties, it 'may equally flow implicitly' from primary or secondary EU law. This principle has been developed by a long list of cases following the judgment of the Court of Justice in the *ERTA* case[21] and is nowadays enshrined in Article 3(2) TFEU. According to that provision, the EU enjoys exclusive competence to negotiate an agreement where the conclusion of an international agreement 'is likely to affect common rules or alter their scope',[22] where the area of EU rules that risks being affected or altered has to be 'already covered to a large extent'.[23] Indeed, exclusive competence

[21] Case C-22/70 *Commission v Council* (European Agreement of Road Trasport – ERTA), ECLI: EU:C:1971:32.

[22] Case C-600/14 *Germany v Council* (COTIF), ECLI:EU:C:2017:935, paragraph 49.

[23] *Opinion 3/15* (Marrakesh Treaty to Facilitate Access to Published Works for Persons who are Blind, Visually Impaired or Otherwise Print Disabled), ECLI:EU:C:2017:114, paragraph 106 and case law cited.

may apply even to likely *future* EU rules, as 'the foreseeable future development of Union law' in that area.[24] To ensure that EU Member States do not accept overlapping or even contradicting commitments, it would appear logical that the EU should participate in negotiations – but this is ultimately a policy choice that the EU Member States must make, taking into account all the political considerations.

D. The Two Levels of Rule-Making

As is the case in all similar situations, the policy choice regarding engagement has an important legal dimension. A comparison of the Fourth Revised Draft of the LBI with the EU Directive shows that the two instruments have similar aims but take rather different approaches in terms of the scope of obligations and legal technique, reflecting the different methods of standard-setting prevalent in the EU and at the UN level. The main substantial difference between the two is that the EU Directive has detailed provisions on due diligence (self-policing by companies, including by maintaining a complaints mechanism, also supervised by national supervisory authorities) and only one provision on judicial remedies for victims. Conversely, the Fourth Revised Draft has only one generally worded article on due diligence but several prescriptive articles on 'access to justice'. These are in reality a set of provisions imposing very extensive liability, jurisdictional, procedural and applicable law rules, not necessarily connected with due diligence, and aimed at, inter alia, allowing human rights victims to choose the most favourable jurisdiction and law to pursue their compensation claims against companies, including for those arising 'from their business relationships'.

Contradictory obligations envisaged in the two instruments could lead to legal uncertainty and conflicting rules if the adoption of both were pursued without coordination. For example, the key concept of 'human rights abuse' is defined by the Fourth Revised Draft broadly, as 'any acts or omissions that take place in connection with business activities and results in an adverse human rights impact'.[25] In terms of scope, the LBI 'shall apply to all business activities, including business activities of a transnational character',[26] but States Parties

[24] Case C-66/13 *Green Network SpA v Autorità per l'energia elettrica e il gas*, ECLI:EU:C:2014:2399, paragraphs 61–64; *Opinion 1/03* (New Lugano Convention), ECLI:EU:C:2004:490, paragraph 126. See also Case C-114/12 *Commission v Council* (Neighbouring Rights of Broadcasting Organisations), ECLI:EU:C:2014:2151, paragraph 70.

[25] Fourth Revised Draft, Article 1.3.

[26] Fourth Revised Draft, Article 3.1.

may differentiate how business enterprises discharge their obligations, depending on their size, sector and other criteria. By contrast, the EU Directive defines 'adverse human rights impact' by reference to more clearly defined 'abuses of human rights', with the relevant human rights, conventions and instruments listed in the Annex.[27] Moreover, its scope is more precise, as it is meant to extend to EU companies with over 1000 employees and net turnover of at least €450 million, as well as non-EU companies based in the EU that fulfil the same turnover criterion.

The most important obligation in the Fourth Revised Draft requires states to effectively regulate all business activities of enterprises within their jurisdiction, including those that undertake 'activities of a transnational character'.[28] In particular, they must 'adopt appropriate legislative, regulatory, and other measures' to 'prevent the involvement of business enterprises in human rights abuse' and 'ensure respect by business enterprises for internationally recognized human rights and fundamental freedoms'.[29] To that end, they must adopt 'legally enforceable requirements for business enterprises to undertake human rights due diligence as well as such supporting or ancillary measures as may be needed', which must respect certain listed general criteria.[30] The obligations found in the EU Directive are, by comparison, more precise: the latter requires the EU Member States to ensure that companies conduct human rights and environmental due diligence, notably by integrating due diligence into all corporate policies and by 'preventing and mitigating potential adverse impacts, and bringing actual adverse impacts to an end'.[31] They must 'identify and assess actual and potential adverse impacts arising from their own operations or those of their subsidiaries and, where related to their chains of activities, those of their business partners'.[32] Moreover, companies must take appropriate measures to prevent potential adverse impacts identified or to adequately mitigate those impacts, bring to an end actual adverse human rights and environmental impacts and provide for the possibility to submit complaints to the company in case of legitimate concerns regarding those potential or actual adverse impacts, including in the company's value chain.[33] Finally, they must 'periodically assess the implementation of their due diligence measures'.[34]

[27] EU Directive, Article 3(c).
[28] Fourth Revised Draft, Article 6.1.
[29] Fourth Revised Draft, Article 6.2.
[30] Fourth Revised Draft, Article 6.4.
[31] EU Directive, Article 5.
[32] EU Directive, Article 8(1).
[33] EU Directive, Articles 10, 11 and 14.
[34] EU Directive, Article 15.

On the key issue of legal liability, the Fourth Revised Draft provides for an extensive regime, requiring a

> comprehensive and adequate system of legal liability of legal and natural persons conducting business activities, within their territory, jurisdiction, or otherwise under their control, for human rights abuses that may arise from their business activities or relationships, including those of transnational character.[35]

By comparison, the EU Directive sets out the requirement for Member States to lay down rules governing the civil liability of the company for damages arising due to its failure to comply with the prevention and remedy obligations, under specific conditions.[36]

The other substantive area of law covered by the Fourth Revised Draft concerns enhancing access to justice for victims of human rights abuses, with provisions on expanded jurisdiction, choice of law, liability and modification of civil procedure rules. While the EU Directive is focused on due diligence of companies as the core of the envisaged legal mechanism, to be reviewed and sanctioned by public authorities in cases of non-compliance, the Fourth Revised Draft goes further. It includes a series of provisions seeking to ensure the widest opportunities for legal redress for victims, including in the legal systems of the countries where the companies are domiciled. These legal remedies are envisaged as self-standing and functionally independent of the due diligence process. To complicate things further, several of the provisions of the Fourth Revised Draft correlate with several instruments of the EU *acquis* in the area of civil justice cooperation. Conversely, there are also several provisions where there is neither an existing EU *acquis* nor any planned legislation at the level of the EU. These are, in particular, the provisions on evidence and elements of civil procedure – even if the EU Directive contains several principles in that regard[37] – as well as the issue of criminal responsibility of legal entities. Given this complexity, it would not be easy even for a seasoned EU lawyer to determine what exactly the EU can and cannot do at different stages of negotiations.

If the negotiations toward a human rights due diligence LBI continue as they have been conducted thus far, it can be inferred that authority to negotiate provisions covering the entire LBI arguably falls within the sphere of both EU competence (for some parts) and EU Member State competence (for other parts).

[35] Fourth Revised Draft, Article 8.
[36] EU Directive, Article 29.
[37] EU Directive, Article 29(3).

This duality creates a problem both for the representation of the EU in Geneva and in internal EU decision-making in Brussels. In the framework of the LBI deliberations, EU representatives may only speak on issues that fall within its competence, unless they are authorised by the EU Member States to also express positions falling within the Member State competence. On the other hand, the breadth of the LBI poses a significant problem for the EU internally, as its implementation, if the EU were to accede to the instrument as currently envisaged, would likely require legislative changes at both the EU and the Member State level. In view of the changing substance of the envisaged instrument and especially the heated discussions at the eighth and ninth OEIWG sessions, which indicated that any agreement is far from certain, it is not easy to identify the full legal implications and corresponding internal procedures that would be necessary to implement internationally agreed rules in the EU.

Alternatively, depending on the political interest in the EU, the Council could authorise EU institutions – the EEAS and the Commission – to only negotiate in subject areas that correspond to the content of the EU Directive. In this scenario, the EU could seek to align, as far as possible, the LBI with the main principles of the EU Directive on due diligence obligations, align the proposed liability rules in the same sense, and support inclusion of general principles on access to justice rather than any specific rules in the area of judicial remedies. This could be done, for instance, by taking inspiration from Article 47 of the EU Charter of Fundamental Rights.

V. Conclusion

The example of the proposed instrument for human rights due diligence that is being elaborated under the auspices of the HRC provides an insight into several structural issues faced by the EU when it seeks to act internationally and within the UN framework in particular. The internal and the external dimension are closely interrelated, as is evidenced by the difficulties in determining the rights process, competence and content of regulation in an increasingly conflicting global political environment. The difficulties in aligning the internal EU process –both in terms of EU legislation on due diligence for companies and the potential decision to engage in HRC negotiations –with the specific position of the EU in the HRC, demonstrate the difficulties in establishing an efficient rule-making process and the challenges that the EU has to face in pursuing its human rights agenda internationally. It is hoped that if the LBI negotiations ultimately lead to a text that the EU could engage on, this could provide a basis for further institutionalising its role in the HRC and placing it on a more solid footing.

INDEX

www.ingramcontent.com/pod-product-compliance
Ingram Content Group UK Ltd.
Pitfield, Milton Keynes, MK11 3LW, UK
UKHW020819290926
15470UKWH00003B/1263

9 781509 967711